极 地 科 学 丛 书
Polar Science

极地文化遗产

不容有失的宝藏

苏珊·巴尔 著
苏平 译

Polar Cultural Heritage
Too Important to Lose

Susan Barr

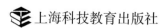
上海科技教育出版社

本书的研究和出版得到中国海洋发展研究会项目（项目编号：CAMAZD201605）和挪威王国驻上海总领事馆的支持。

With the support of the Royal Norwegian Consulate General in Shanghai and the Project of China Association of Marine Affairs (Project No. CAMAZD201605)

作者简介

苏珊·巴尔（Susan Barr），伦敦大学学院斯堪的纳维亚研究荣誉学士，挪威奥斯陆大学历史考古学博士。自1979年以来，她一直专注于极地事务，是挪威北极地区（斯瓦尔巴和扬马延）的第一位全职文化遗产官员，曾担任位于奥斯陆的挪威极地研究所历史和收藏部门负责人，以及挪威文化遗产局的极地遗产管理者。她是国际古迹遗址理事会下设国际极地遗产委员会的创立者兼主席，2014年至2018年任国际北极科学委员会的主席。她的目标之一是提升人们对极地科学中的人文学科的重视程度。她还是奥斯陆弗雷姆博物馆的董事会成员，也是一些其他极地董事会和委员会的成员。她在北极和南极/亚南极地区进行过大规模的实地工作，并撰写了许多有关极地历史和文化遗产的书籍和文章。

译者简介

苏平，同济大学外交学系讲师，硕士生导师，毕业于中国社会科学院世界经济与政治研究所，美国约翰斯·霍普金斯大学尼采国际问题研究中心、挪威南森极地研究所、挪威国际关系研究所访问学者，美国国务院SUSI项目美国政治与外交类访问学者。主要研究领域为国际关系理论、全球治理、国际组织、北极。苏平是国际问题研究协会（ISA）会员、国际北极科学委员会社会与人文工作组中国代表（IASC）、国际北极社会科学协会会员。主持并参与国家级、省部级课题数10项，发表SSCI、CSSCI期刊论文10多篇，译著1部，获上海市哲学社会科学优秀成果三等奖。

Author biography

Susan Barr has a BA-Honours degree in Scandinavian Studies from University College, London and a PhD in historical archaeology from Oslo University, Norway. She has worked solely with polar matters since 1979, as the first full-time cultural heritage officer for the Norwegian Arctic (Svalbard and Jan Mayen), as head of section for history and collections at the Norwegian Polar Institute in Oslo, and as director for polar heritage at the Norwegian Directorate for Cultural Heritage. She was the Founding President of the International Polar Heritage Committee within ICOMOS—the International Council on Monuments and Sites—and was President of IASC (the International Arctic Science Committee) 2014–2018 when one of her aims was to raise the status of the Humanities in polar science. She is also Board Member of the Fram Museum in Oslo and member of other polar boards and committees. She has had extensive field work in the Arctic and Antarctic/sub-Antarctic and is the author of books and articles concerning polar history and cultural heritage.

序言

从远古蛮荒到现代文明，人类经历了漫长的发展时期，也产生了许多具有历史意义的文化遗产，从精美的宫殿、寺庙、石窟壁画到工业遗址。文化遗产不仅提供历史信息，还承载着人类精神文明的宝贵财富，它的保存与保护反映了人类的文明程度。然而文化遗产正受到来自各方的挑战，一些历史文化遗产因为没有得到及时认定而被作为废墟拆除，一些历史文化遗产则因其广为人知的名气而被"爱到窒息"，比如潮水般涌到威尼斯和希腊岛屿的游客对文化遗产的保护造成了巨大压力。

极地是世界文化遗产保护的重要地区之一。人们常常为极地的神秘、辽远、宁静而着迷，其极致纯粹的自然环境和广袤无际的原始荒凉更使人充满探索的好奇。然而，在壮丽的自然之外，人类不畏艰险、充满智慧和勇气探索未知的壮举在南北极大地上留下了众多重要的人文历史遗迹。追溯并保护这些文化遗迹，不但可以重温历史，保存宝贵的人类活动记录，更可使人类探索未知的坚毅等蕴含其中的宝贵精神财富得到传承。然而，文化遗产不像生物，后者可以作出适应气候与环境改变的调整，而前者一旦消失就永不复还。随着越来越多的游客和游轮到达南北极，极地文化遗产被人们无意或有意的行为毁掉的风险日益增大，极地文化遗产保护的任务艰巨而又紧迫。

苏珊·巴尔博士毕业于伦敦大学学院和奥斯陆大学，获历史考古学博士学位，自1979年起，一直从事极地历史和文化遗产研究和保护工作。1988—2016年，她曾担任挪威文化遗产局极地遗产管理者和挪威研究理事会极地科学委员会委员。苏珊有着丰富的南极和北极实地工作经验，出版了一系列关于极地历史和遗产的专著，她是国际古迹遗址理事会（ICOMOS）国际极地遗产委员会的创立者兼主席。

苏珊曾担任国际北极科学委员会（IASC）主席，非常有幸，我作为副主席曾与她在IASC执行委员会共事了6年。IASC的宗旨在于鼓励和促进北极研究的所有方面、从事北极研究的所有国家和在北极地区的所有领域之间的合作，促进和支持跨学科研究，增进对北极及其在地球系统中作用的科学理解。在苏珊担任主席期间（2014—2018），IASC 制定了北极研究的战略计划，通过促进北极研究合作、促进参与以及确保知识交流三大支柱，围绕北极在全球系统中的作用、观测和预测北极未来的气候动力学与生态系统响应以及理解北极环境与社会的脆弱性和恢复力、支持北极可持续发展三个优先领域，推进国际北极研究。

2017 年 10 月，苏珊·巴尔应邀出席了在吉林大学召开的中国极地科学学术年会，在长春和上海分别与中国极地研究学者和社会公众交流了气候变化对极地文化遗产的各种影响和国际社会对极地文化遗产的保护行动。博克格雷温克在南极阿达尔角建造的小屋是人类在南极洲建造的第一个建筑，根据新西兰和英国向南极条约协商国会议的建议，该小屋以及特拉诺瓦小屋的遗骸被认定为历史遗迹或纪念碑（HSM 22）。同年出发的中国第 34 次南极考察队，在 2018 年 2 月随"雪龙号"极地考察船驶往南极阿达尔角，派出了一架 Ka-32 直升机和 40 名南极考察队员，与 4 名新西兰南极考察队员合作，共同实施了保护博克格雷温克小屋的修复任务，积极履行了我国保护南极文化遗产的应尽责任。

每一次的失去，都将是永恒的消失，南北极文化遗产的保护，已经刻不容缓。这是一本苏珊为中国读者专门著写的书，由同济大学的极地问题研究青年学者苏平博士翻译。希望本书的出版，能够唤起大家更多的参与和合作、更加自觉的意识和行动，去预防和阻止自然侵蚀和人为破坏的发生，更加负责和科学地保护好南北极的文化遗产。

中国极地研究中心主任
国际北极科学委员会副主席
杨惠根
2019 年 4 月

目 录

- 001 —— 引言
- 003 —— 自然与文化
- 007 —— "文化"及文化遗产的含义或定义
- 009 —— 文化遗产的价值
- 015 —— 极地地区
- 019 —— 原住民历史
- 020 —— 外来移民历史
- 040 —— 一些标志性的极地探险
- 050 —— 南极海豹猎人和早期因纽特人的遗址
- 053 —— 早期文化和活动的物质证据
- 059 —— 木屋的意义远大于其本身
- 062 —— 《南极历史遗址及古迹名录》
- 067 —— 极地自身的文化遗产
- 071 —— 我们还有"极地荒原"吗
- 075 —— 极地文化遗产不可否认的巨大吸引力
- 081 —— 对北极文化遗产的威胁
- 087 —— 文化遗产的负面影响有助于解释气候变化
- 092 —— 用科技保护文化遗产
- 096 —— 怎样减少旅游业的影响
- 099 —— 对于未来的考量
- 102 —— 注释
- 104 —— 图片来源

Contents

- 109 — Introduction
- 111 — Nature and Culture
- 116 — The meaning or definition of "culture" and cultural heritage
- 118 — Cultural heritage values
- 125 — The Polar Regions
- 128 — Indigenous history
- 129 — Non-indigenous history
- 149 — Some iconic polar expeditions
- 158 — Antarctic sealers and early Inuit sites
- 161 — The material evidence of earlier cultures and events
- 167 — A hut is more than just a hut
- 170 — The Antarctic Historic Sites and Monuments list
- 175 — Cultural heritage at the Poles themselves?
- 179 — Do we still have "polar wildernesses"?
- 183 — The undeniable attraction of polar cultural heritage
- 189 — Threats to the Arctic's cultural heritage
- 195 — Negative effects on cultural heritage help to illustrate climate change
- 200 — Using technology to help protect cultural heritage
- 204 — How to lessen the impact of tourism?
- 207 — Thoughts about the future
- 210 — Notes
- 212 — Picture Sources

引言

图1 荒凉而美丽的极地景观,冰雪当然是其特色,南极企鹅、北极熊和海象则是其标志。南北两极拥有大量文化遗产古迹和遗址,可以帮助我们更好地理解人类与极端环境的历史互动。失去这些历史参考坐标,世界将会变得了无生趣。

▼

极地地区——北极高纬度地区和南极地区——既是许多人心中遥远的所在,也是当前全球气候变化讨论的中心。虽然目前对气候变化究竟是自然演化的结果还是人类活动影响所致尚未有定论,但对于世界各地的人来说,现在的气候明显不再是他们所习惯的气候,也不再像以前那么稳定。这一变化在北极地区体现得最为明显,全球变暖正在改变这个迷人地区的自然和生态。北极是一片被冰雪覆盖的海洋,四周被陆地包围,但过去几年里,海冰覆盖范围在急剧减少,厚度也在迅速降低。北冰洋的开放导致并加剧了天气模式的变化，进而影响了南部地区。南极是一片被冰雪覆盖的大陆，四周被海洋环绕，虽然变暖的影响目前还不是那么明显，但是确实存在，也将影响到我们所有人。

北极原住民现在已经无法再遵循祖祖辈辈留下来的传统，他们不得不适应新的生活方式。包括阿拉斯加北部海岸在内，很多地区不再有冰雪覆盖，对当地原住民来说，冬春两季在海冰上的狩猎之景正逐渐变成回忆。格陵兰岛等地区的冰面也不再安全，猎人们都是冒着生命危险在狩猎，而在此之前，脚下厚而坚固的冰总能让他们心安。各种走兽、鸟类和鱼类的数量也发生了变化。受北极气候变暖变潮的影响，即使是现代化的人类定居点和交通基础设施也受到了威胁。

图2　格陵兰岛上的因纽特族儿童在学习驾狗，摄于2018年。

关于壮美的两极地区，可讲的故事有很多。然而，本书探讨的重点是两极地区的文化遗产。在进行具体探讨之前，本书将首先剖析"文化遗产"一词的含义。

自然与文化

▼

最初,世界上只有大自然。不过,在人类出现以后,他们便开始用工具与文化对之前从未受人类影响的原始自然进行改造。与先于人类出现的其他动物类似,早期的人类通过狩猎和采集食物来维持自身生存。在开始运用智慧改造周围环境之前,人类对自然和生态的影响几乎可以忽略不计。然而,当人类开始使用先进工具和方法为直系亲属、族群、部落乃至其城市和国家提升食物采集规模和效率之后,一切都发生了变化。科学研究尚未发现改变人类已知史前史的具体证据,因此,人类开始使用工具的确切时间仍不确定,但2015年在肯尼亚发现的原始工具将人类开始使用工具的时间从两百五十万年前提前到了三百多万年前[1]。

目前,学界认为,我们人类,也就是智人(*Homo sapiens*),大约是在31.5万年前在非洲进化产生的[2],大约1.7万至1.2万年前,为了满足自身需求,人类开始发展农耕和农业文化、并尝试对动植物进行改造,从此彻底改变了人类和环境的互动方式。焚烧植被和拓荒耕种改变了当地生态系统。人类开始开辟之前以森林和植被为主的区域。此类活动彻底改变了自然环境,某些动植物得以更好地繁衍生长,另一些则退化消失。人类驯化动物以便更容易地获取肉、奶、毛皮和其他物品,植物生态学和新的土地使用方法改变了自然环境。为了换取食物和安全,曾经的野生动物成了人类狩猎的帮手和盟友,比如说狗和猛禽。

工业化之前,人类仍然成功地保持了与自然的合理平衡。随着人类采取定居为主的生活方式,园艺、农业、动物养殖、捕鱼

图 3　乍得的一个传统非洲村庄，摄于 1973 年。

和食品加工及储存技术得到了发展，更多人能终生聚居一处。

　　一旦某些定居点能够被组织起来，使其粮食生产效率提高至足以将部分社会活动从狩猎和农耕等日常工作中分离出来并进行分工的程度，便产生了我们今天所谓的文明。此时，制定群体行为规范和法律、成立管理部门、建造专属建筑物、组织防御系统并制定维持惯例的书面文件就成了必需。我们今天知道的伟大历史文明都起源于农业，尤其是大河流域的农业。河流不仅提供了文明内部必需的淡水，还是运输和贸易要道。人类对水的控制催生了绘画和书写。绘画和书写提高了人类的定居水平，并由此衍生出人类的首批主要文明。大约 5 200 年前，在今天的伊拉克南部，即幼发拉底河和底格里斯河在非洲东北部和亚洲西南部的交汇处，诞生了人类最早的古代文明。随后在尼罗河沿岸即今天的埃及产生了另一文明。大约在 4 500 年前，同样的事情发生在印度的印度河沿岸。4 000 年前，克里特岛诞生了爱琴海（地中海）文化。3 600 年前，中国诞生了商文化[3]——不过，在此之前中国就已经拥有了接近一百万年的不间断史前史。中国北方（黄河以北)有着比其他任何地区更为连续的人类发展史[4],其历史可以追溯到所谓的"北京人"。这些在北京周口店附近发现的人类牙齿和已灭绝的直立人的骨头可追溯到77万至23万年前[5]。美洲最早的文明诞生在3 200年前的中美洲和安第斯山脉。

尽管有上述及后来的人口中心聚居点存在,但是直至工业化出现人类才完全具备了改变自然环境的能力。工业化之前,人类用手工工具进行农业生产,用动物或人拉犁,用袋子和水桶运水,现在则是在万顷粮田上进行机械化收割。

图4 工业化小麦收获现场。

今天,我们改造河流、建造巨大的水坝形成人工湖,湖水淹没了大量村庄和农田。我们在自然植被上喷洒化学药剂,消灭帮助作物授粉的昆虫,还将大片土地置于混凝土之下。为了种植大豆和油棕榈等单一作物,我们摧毁雨林,这些雨林却再也无法从改造后的废土中再生。为了满足眼前的利益,我们对动植物进行基因改造。我们正在一步一步将原始自然从地球上抹去。就连海洋也成了我们倾倒垃圾、排泄污物和丢弃塑料的场所。这是人类文化对自然的最终"胜利"吗?

稍后我们将看到人类与自然的互动如何影响了极地地区。

图 5
a. 婆罗洲,因伐木而遭毁坏的雨林。
b. 海滩上的塑料污染。

"文化"及文化遗产的含义或定义

▼

语源学(即研究词的起源和历史的科学)从两个视角定义了"文化"一词：哲学与逻辑的角度，农业、狩猎与渔业的角度[6]。从哲学与逻辑的角度，文化是指整个社会或群体的智力活动和物质生产活动(如石器时代文化、维京时代文化)或遍及整个群体的集体社会行为、价值观和规范。从农业、狩猎和渔业的角度，文化则被定义为对土地、森林、植物或水的开发、利用。这使我们不得不再次面对原始自然与人类文化之间的冲突，后者包括工具、信仰、艺术和文学在内的物质生产结果和人类活动。

联合国教科文组织用下列方式界定了"文化遗产"一词：

物质文化遗产包括：

- 可移动文化遗产(绘画、雕塑、钱币、手稿)
- 不可移动文化遗产(古迹、考古遗址等)
- 水下文化遗产(沉船、水下遗址和城市)

非物质文化遗产包括：口头传统、表演艺术、仪式[7]。

可以说，经由人类制造或改造的任何东西都是文化遗产，也就是说，决定个人或特定时期文化的是文化物品而不是自然物品。即使是人类在森林中或在苔原上留下的足迹也是人类活动的证据，因此也与人类文化和人类行为有关。在研究当时的食品类型、食品保存、饮品、产酒区和非产酒区之间的贸易、口腔卫生习惯等时，空食品罐、酒瓶或者使用过的牙刷都可以为我们提供一些线索。根据物品的年代和周边环境，我们可能会将这些物品视为垃圾或有趣的研究对象。如果研究对象是在100多年前废弃在丛林深处的营地或20世纪初的南极洲小木屋里发现的，那么这绝对比在现代城镇的垃圾箱中的发现有更大的历史意义。因此，在被视为值得保护和研究的文化遗产与被视为一次性垃圾的物品之间存在一定分界线，但非专业人员很难清楚地理解这种区别。

二者的区别在极地地区尤其重要，下面将作进一步的解释。

图6 瑞典某科学探险队在斯瓦尔巴悲伤峡湾（冷岸群岛）的营地遗址，该探险队1899—1900年间在此做有关地球形状的测量。图片左下角可以看到由玻璃瓶堆成的垃圾堆。整个遗址，包括这个垃圾堆，都受到挪威斯瓦尔巴群岛环境法的保护。

文化遗产的价值

文化遗产被赋予的价值有助于确定其是否值得保护。但不幸的是，文化遗产的价值是无法用数学计算的（尽管有些文化遗产的价值可能完全由其年代决定），而是取决于专业文化遗产工作者的专业知识。文化遗产的价值可能随时间变化而升级或降级。例如，挪威的第二次世界大战遗迹原本应该在战后的几十年内清除和摧毁，但在前几年，这些遗迹被认定为值得保护的文化遗产，虽然这些遗迹会勾起人们极为痛苦的回忆。世界各地对文化遗产价值的看法也不同。在某些地区被认为有碍观瞻而必须拆除的废弃工业用地，在其他地区可能会被视为地方、国家甚至国际遗产而受到保护。联合国教科文组织的世界遗产名录既包括精美的宫殿和寺庙，也包括工业遗址。

文化遗产可以作为单一物体受到法律保护，例如矗立在田野中的维京时代晚期的单个符文石，或者具有历史意义的南极探险队留在南极岩石中的贮藏物。然而，目前的趋势是尽可能将所有

图7 中国明清两代的皇宫（a）和德国的采煤工业园区（b）均被联合国教科文组织列为世界文化遗产。

有助于解释研究对象产生的历史情境的周边环境都包括在内,因为在某些情况下,如果脱离周边环境,就很难准确理解研究对象。一个很好的例子就是在南极大陆和半岛存在许多极具历史意义的小屋,探险家和科学家曾在这些小屋中度过南极冬天极端恶劣的天气,如果将其移到其他大陆的博物馆,那么一个摇摇欲坠的小木屋与南极极端恶劣环境之间那种引人深思的强烈对比感就会消失。因此,博物馆或其他新的容纳地必须进行创新和创造,比如通过环绕电影或模拟南极寒冷多风的环境,帮助游客身临其境地

图 8
a. 南极洲阿达尔角的历史小屋。
b. 位于瑞典的符文石。

"看到和感受到"南极极端环境中的小木屋。

因此，所有文化遗产都可以为我们提供当前和过往社会的历史信息。特别是在有文字记录的历史出现之前，人类活动的痕迹可以增进我们对祖先的了解，进而有助于解释我们如何成为今天的我们。从考古发现的早期工具一直到今天的工业化机械，可以看到几千年来动物驯化和农业发展的历程。在考古未能发现早期工具的地方，岩石艺术和洞穴绘画可以帮助我们解读人类很久以前的活动。

阿尔及利亚的阿杰尔高原联合国教科文组织世界遗产地内有15 000件岩画作品，其历史可追溯到12 000年前。这些岩画描绘了人类放牧牛群和狩猎的场景，还描绘了羚羊和鳄鱼等野生动物。甚至像之前提到的空食品罐这样平淡无奇的物品也可以帮助解释历史。例如1872年至1873年间，一队本不打算越冬的船员在挪

图9　位于印度的比莫贝特卡石窟，已被列入联合国教科文组织世界文化遗产名录。岩棚内的画作描绘了早期人类的生活场景，其中有些岩画创作于约9 000年前的石器时代。

威北极高纬度地区斯瓦尔巴群岛留下了一些食品罐。最近的调查显示，用于密封罐头的铅是导致他们死亡的重要原因，他们是在吃了这种罐头食物后，因铅中毒而死亡的。

图10　1872年至1873年间，17名船员在斯瓦尔巴群岛的一所房子里过冬并留下了一些空食品罐。尽管屋内住宿条件良好且食物储备充足，但所有人都死了。一直以来，人们都认为他们死于坏血病（饮食中缺乏维生素C所致），但最近的研究表明，用于密封罐头的铅条引发了铅中毒，最终导致他们死亡。

　　随着研究方法的改进，我们能从新老文化遗产中获取的信息越来越多。这也许可以帮助我们更好地适应未来，例如，传统建筑方法将指引我们找到更环保、更节能的未来房屋建筑方式。

　　对大多数人来说，文化遗产的主要价值不一定是其历史研究价值，而是"亲身"体验（特别是在文化遗产原址体验）所带来的价值。北极高纬度地区异乡人孤独的坟墓，可以增进到访的当代游客对17世纪因坏血病、疾病或意外死亡的欧洲猎人的理解。前提是游客要用心对几百年前发生的事情进行体悟，而不是仅仅在匆忙赶往下一景点之前赶紧拍几张照片完事。

　　即使很小的物品也可以让人们产生无限联想，比如小屋墙壁

图 11　今天，在斯瓦尔巴群岛可以看到许多这样的简陋坟墓。这些是 17、18 世纪捕鲸者的坟墓，他们趁着北极短暂的夏季从欧洲航行至此进行捕鲸活动。由于永久冻土过于坚硬，坟墓只能挖得很浅，所以挖墓时，人们会在棺材上覆上大石头，以防止北极熊和狐狸打扰尸体。

图 12　加拿大东北部一个萨卡克文化住宅遗址，这一遗址对研究大约 4 500 年前定居格陵兰的加拿大人极其重要。照片显示的石头围作的圆圈是皮制帐篷的边界，中间石头围作的方圈则是炉子的位置。

上的老照片、破损的铁锹、鲸椎骨做的凳子。苔原上一圈不显眼的石头，也许中间的几块是用来造炉子的，这一切都可能意味着曾经有一个因纽特人家庭在这个我们不愿短暂野营的地方住了或长或短的时间。

能看，但是不能触摸！对绝大多数人来说，要想在原始状态下体验地道的文化遗产，乘船或巴士做短暂观光或参团旅游是他们所能做到的最大限度，当然，此类体验可以帮助我们理解相关文物或文化遗址为什么需要受到保护以及如何对其进行保护。但为帮助游客进行游览而设置的标志和障碍物无疑会削弱游客的敬畏之心（尽管必然要设置）。时至今日，只有当有幸在中国长城的某一段城墙上独自走一走，或者在没有旅游团的初冬早晨在雅典卫城上站一站时，我们才有可能与历史进行有意义的交流。但很遗憾，对于大多数人来说，这现在是几乎不可能的经历。

图 13　希腊岛屿海滩上的游客。大量游客涌入可能会产生问题，比如著名的意大利城市威尼斯正深受其名气带来的不利影响。威尼斯以其星罗棋布的运河和水道而闻名，每天要接待 60 000 名游客，一年则有超过 2 000 万游客。潮水般涌入的游客逼得当地人不得不搬到清净点的地方居住，那些最受欢迎的景点更是不堪其扰。威尼斯和这张照片所示只是众多被"爱到窒息"的历史遗迹中的两个，当地政府正考虑如何控制游客数量。

极地地区

"看到一只北极熊便知道你在北极，看到一只企鹅就知道你在南极"。这种区别南极和北极的俏皮方式是正确的，当然这只是对两个区域作了最浅显的区分。相对来说，南极洲更容易定义。自1959年12月1日《南极条约》签订，并于1961年6月23日生效以来，南极地区被定义为南纬60°以南的陆地和海域。南极大陆总面积达1 400万平方千米（几乎是澳大利亚面积的两倍），其中约有98%被冰覆盖，冰层平均厚度为1.6千米，存储有世界上大约70%的淡水资源。冬季来临时，周围海水结冰，冰层从海岸向外延伸约1 000千米，冰封区的面积几乎翻倍。南极洲大陆存在于南纬70°以南，也是通常描述南极极端气候时所指的区域。然而，作为南极"手臂"的南极半岛连同其他岛屿一直从南极大陆的西北部几乎延伸到了60°纬线。因此，南极半岛及岛屿区最容易乘船抵达，且与南极主大陆相比，这些地区的气候相对温和，因此这里吸引的海豹猎人、捕鲸人、科学家和游客数量都超过了冰川覆盖的南极主大陆。

南极洲从未有过原住民，今天留在那里的人要么是科学家，要么是协助运营科研基地的后勤人员。南极洲即使环境稍微好一点的地方，也根本没有陆地哺乳动物，植被也极少，只有一些地衣和苔藓。夏季，海洋里生活着各种各样的海洋生物，从大型鲸鱼到磷虾乃至微生物，每一种生物都在整个生态系统中发挥着不可或缺的作用。南极的夏天有很多海鸟，尤其是企鹅，但冬天只有雄性帝企鹅留在南极大陆，它们聚集在一起，各自孵化身下的一枚蛋。已经确定的是，人类在南极洲首次出现至少可以追溯到1820年，当时的海豹猎人发现南极有大量各类海豹，随后开始大肆捕杀。虽然有12个国家宣称在南极洲拥有领土，但1959年签订的《南极条约》搁置了这些国家的领土要求，各签署国以协商一致的方式引入了国际政府的概念。目前有53个国家签署了该条约，该条约的大前提是南极大陆仅用于和平目的以及科学考察与合作。

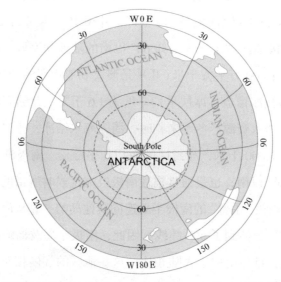

图14 南极洲地图。（图中文字：South Pole，南极点；Antarctica，南极洲；Atlantic Ocean，大西洋；Indian Ocean，印度洋；Pacific Ocean，太平洋）

北极地区相对来说更难界定，因为它是多国陆地包围的海洋，这些国家向南延伸的范围超出了任何有关"北极"的界定范围。五个国家的海岸线直接与北冰洋接壤：挪威、丹麦（格陵兰）、加拿大、美国和俄罗斯，冰岛、瑞典和芬兰则包括了北极圈以北的陆地区域，即大约北纬66°以北。然而，如果按照其他界定方法，有些地区并不属于北极，例如挪威西北海岸，该海岸的温度由温暖的墨西哥湾流的支流调节。挪威研究理事会将挪威北极领土定

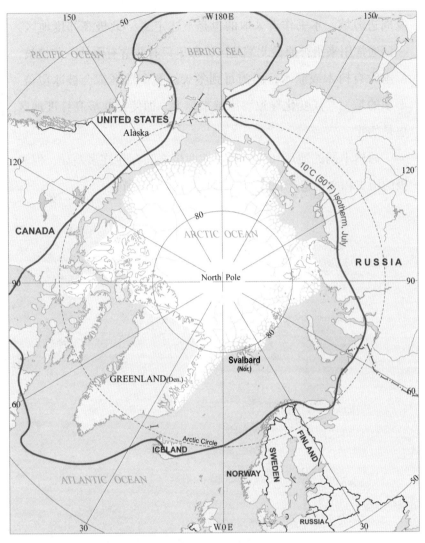

图15 北极地图。虚线圈是北极圈，红色的是7月的10℃平均等温线。[图中主要文字：Svalbard（NORWAY），斯瓦尔巴群岛（挪威）；ICELAND，冰岛；SWEDEN，瑞典；Greenland（DENMARK），格陵兰（丹麦）；FINLAND，芬兰；RUSSIA，俄罗斯；Alaska（UNITED STATES），阿拉斯加（美国）；CANADA，加拿大；North Pole，北极点；Arctic Ocean，北冰洋；North Atlantic Ocean，北大西洋；North Pacific Ocean，北太平洋；Bering Sea，白令海]

义限定为挪威大陆以北的领土：即斯瓦尔巴群岛和扬马延火山岛及周边海域。本书作者采取的也是上述定义。"北极高纬度地区"一词通常用来指苔原最北端地区，那里只有非常有限的一些植被，根本没有树木或灌木，因为每到冬天，整个地区都会被冰覆盖。不幸的是，气候变化导致海冰越来越少，即使是北极高纬度地区也是如此。

迥然不同的地理和政治环境使两极之间存在许多差异，但两极的历史和文化遗产有许多相似之处。如后文所述。

原住民历史

极地区域人类活动的历史有新有旧。早在28 000年前，人类就开始从东亚较温和的地区迁移到西伯利亚东北部。在距今110 000—11 700年前的更新世（或称冰川期）的最后一个冰期末期，人们开始越过西伯利亚和美洲西北部之间的白令陆桥，即后来的白令海峡。当时，一小部分人定居在白令陆桥和现在的阿拉斯加（东白令），但是直到大约12 000年前，大型冰盖仍然阻挡着他们深入北美洲的去路。然而，根据坐落在美洲冰川南部距今15 000—14 000年前的考古遗址，大多数考古学家认为，第一批移民是沿陆桥南岸和美洲西海岸乘船抵达美洲的。大约5 000年前，古爱斯基摩人从西伯利亚移居阿拉斯加，并迅速进入南阿拉斯加，北极小工具传统（登比文化）随之产生。这些古爱斯基摩人以及后来的爱斯基摩人、阿留申人和纳德内人发展成为我们所认识的美洲原住民和第一民族部落。随着加拿大北极地区的冰雪逐渐消融，古爱斯基摩人开始进入这个地区，在4 500年前到达了格陵兰岛和拉布拉多半岛。早期的格陵兰文化被称为萨卡克文化，在加拿大被称为前多赛特文化。到了2 500年前，早期的古爱斯基摩人发展出多塞特文化（或称"图尼伊特人文化"）。多赛特文化始终是北极东部地区的主流文化，直到公元1300年图勒文化出现为止。图勒文化起源于白令海峡地区的新爱斯基摩人猎鲸文化。图勒文化和图勒人就是现今加拿大北部和格陵兰的因纽特人的祖先。这些早期居民的渊源是根据他们的考古遗迹推断而来。[8]原住民面临的生活条件异常恶劣，所以他们的人口一直有限，也没有发展出大规模定居群体。

外来移民历史

在北极人类历史中扮演重要角色的还有外来游客和南方移民。其中最早到来的就是探险家们，他们在北极人类史中占据了重要一席。探险家们热衷于研究北极的地理环境并记录其自然状况。其次是北极自然资源开发商和交易商，他们开发和交易的自然资源包括鲸、海象、海豹、北极熊等海洋哺乳动物，北极狐和驯鹿等陆地哺乳动物，以及煤炭和黄金等矿产资源。对新航线的探索早在15世纪和16世纪就开始了，即分别经北美洲和欧亚大陆北部沿岸的航道，西方人将其称作西北航道和东北航道。此前，早在11世纪，来自俄罗斯的定居者和商人就已经探索了东北航道的

图16 北方海航道近欧亚大陆最北点的切柳斯金角。这张照片拍摄于1990年8月，当时连破冰船在此航线上行驶时也遇到了麻烦。如今，北极冰川的数量要少得多，越来越多类型的船只能够通过这一海上航道。

部分线路。历史上探索新航线的首要驱动力是希望找到从欧洲到东方的新路线，从而在东方找到重要的贸易商品，尤其是丝绸和香料。今天，西伯利亚北部的航线被称为北方海航道。受气候变暖影响，欧亚大陆北部海岸的冰川状况有所改变，作为未来亚洲和欧洲之间的航线，北方海航道将再次引起人们的密切关注。

16世纪的欧洲人认为，在亚洲东部有一条阿尼安海峡将亚洲与美洲隔开。"阿尼安"这个名字应该来自1559年版马可·波罗（Marco Polo）的游记中的一个中国地名"Ania"。意大利商人和探险家马可·波罗详细描述了他13世纪末在中国游历24年的历程。阿尼安海峡出现在16世纪60年代的地图上，也出现在1559年版马可·波罗的游记中。这些地图显示，狭窄而弯曲的阿尼安海峡将亚洲与美洲分隔开来。在欧洲人的想象中，这条海峡是连接欧洲和中国北方的便捷海道。

在11世纪挪威人发现文兰之后，最早在北美北部沿岸进行探索的欧洲人是约翰·卡伯特（John Cabot），1497年，他在英王亨利七世的委托下出海。从那时起，直到1903—1905年挪威探险家罗阿尔·阿蒙森（Roald Amundsen）首次乘船通过航道，探索新航线的历史充满了悲惨结局和英雄事迹。其中最著名的是1845年由约翰·富兰克林爵士（Sir John Franklin）率领的英国探险队，这支探险队的129人和2艘船只失踪。在随后的十年中，有超过40支探险队前往搜寻失踪的船只和人员，其探险的细节不断被发现。就目前所知，探险队船只于1846年夏在航道东侧的以其命名的富兰克林海峡被冰所困，所有人因坏血病和饥饿而丧命，其中许多人死在了向南求助的途中。直到2014年和2016年，世人才在富兰克林海峡以东的威廉王岛附近海底发现了富兰克林探险队所乘的"幽冥号"及"惊恐号"沉船。

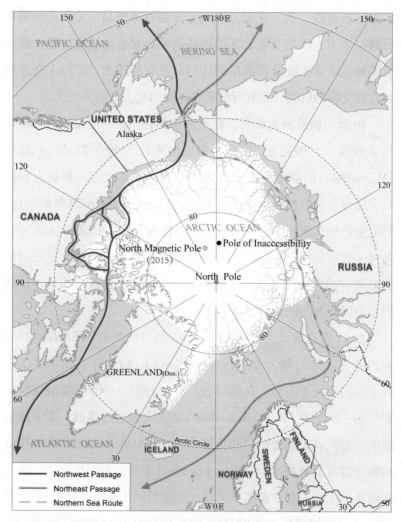

图17　北美大陆以北的西北航道(红色)，欧亚大陆以北的东北航道(绿色)，现在称为北方海航道(浅绿色虚线)。[图中主要文字：PACIFIC OCEAN，太平洋；BERING SEA，白令海；ATLANTIC OCEAN，大西洋；ARCTIC OCEAN，北冰洋；UNITED STATES(Alaska)，美国(阿拉斯加)；CANADA，加拿大；ICELAND，冰岛；SWEDEN，瑞典；GREENLAND(Den.)，格陵兰(丹麦)；FINLAND，芬兰；RUSSIA，俄罗斯；North Pole，北极点；North Magnetic Pole，磁北极；Pole of Inaccessibility，难近冰极]

图18　1895年的一幅著名代表性画作《他们用生命结下最后的联系》，描绘约翰·富兰克林爵士1845年带领探险队在北美大陆以北的西北航道寻找航行路线的命运。他的两艘船被困在加拿大东北部的冰层中，他和128名船员全部丧生。威廉·托马斯·史密斯（William Thomas Smith）根据揭示探险队命运的证据创作了这幅画。

北极高纬度地区文化遗产多样性的典范——斯瓦尔巴群岛

斯瓦尔巴是挪威位于北极地区的群岛，在挪威北部大陆以北，其范围几乎延伸到北纬80°。如今，有人依然会使用其历史名称斯匹次卑尔根群岛，而斯匹次卑尔根岛只是该群岛中最大的岛屿。虽然该岛屿在当前气候变暖期前，一年中的无冰季节只有2—3个月，但是因为受到墨西哥湾暖流流经斯瓦尔巴群岛西部的影响，在该区域向北航行比在北极其他地区更容易。斯瓦尔巴群岛从未有过原住民，但自从1596年荷兰探险队首次发现该群岛之后，其自然资源就不断为临时外来者所开发利用。

在斯瓦尔巴群岛逐渐为人们熟知之际，欧洲商业捕鲸的重要性日益增加。鲸油、鲸须成为广受欢迎的产品。鲸油用来照明、制造润滑油和肥皂，或者用于绳索和衣服的制作过程，鲸须则用来制作马车弹簧、雨伞辐条或者女士束腰。欧洲的北极捕鲸于16世纪末在加拿大东部沿海的拉布拉多-纽芬兰地区开始，并于17

世纪初传播到斯瓦尔巴群岛。特别是来自法国和西班牙交界处巴斯克地区的捕鲸者（巴斯克位于比利牛斯山西端和比斯开湾的南部），他们将传统捕鲸业活动越过大西洋带到纽芬兰地区。

由于掌握了专业的捕鲸知识，巴斯克人常被雇用参与17世纪早期英国和荷兰的斯瓦尔巴群岛捕鲸远征。在众多鲸中，露脊鲸

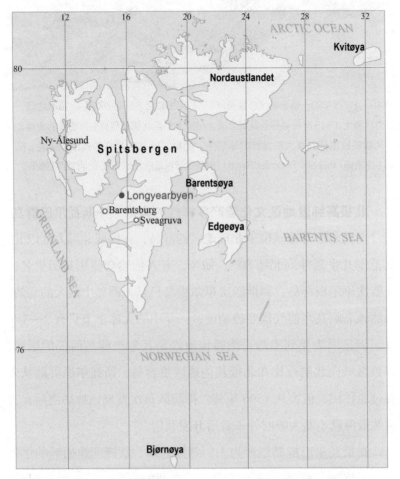

图19　挪威北极地区斯瓦尔巴群岛地图。（图中主要文字：GREENLAND SEA，格陵兰海；NORWEGIAN SEA，挪威海；BARENTS SEA，巴伦支海；ARCTIC OCEAN，北冰洋；Spitsbergen，斯匹次卑尔根岛；Nordaustlandet，东北地岛；Barentsøya，巴伦支岛；Edgeøya，埃季岛；Longyearbyen，朗伊尔城；Ny-Ålesund，奥勒松；Barentsburg，巴伦支堡；Sveagruva，斯维格鲁瓦；Bjørnøya，熊岛）

相对容易捕获。捕鲸人在适宜的海滩上建了捕鲸站,以便从鲸脂中熬煮出鲸油,其中最著名的是斯瓦尔巴西北部的斯米伦堡。斯米伦堡是一个荷兰捕鲸站,被称作"鲸油小镇",在17世纪上半叶建成使用,站内用来熬炼鲸油的锅炉遗址成了热门景点,这里同时也是一个受保护等级很高的文化遗产地。捕鲸是门危险的行当,最重要的原因是捕鲸者必须驾驶小艇接近鲸才能用鱼叉捕杀它们。每年夏季的捕鲸季节,意外、溺水、疾病等致使数百名捕鲸者命丧黄泉。他们的坟墓很浅,因为永久冻土太硬很难深挖。棺材上面盖着大堆的石头。这些石头也只能防一防北极熊和狐狸打扰尸体。

17世纪末期,斯瓦尔巴群岛沿岸再无鲸可捕,该地的捕鲸活动至此告一段落。人们不再在陆地上建造捕鲸站,但是捕鲸活动进一步扩展到了海面。

从18世纪末期开始,冬季探险队来到斯瓦尔巴猎取北极狐和

图20　加拿大拉布拉多红湾,16世纪末—17世纪初的巴斯克捕鲸场遗址,已被联合国教科文组织列为世界文化遗产。遗址中存有从鲸脂中汲取鲸油的遗迹、140名捕鲸者的坟场以及在海湾沉没的大型和小型捕鲸船。

图21 该幅18世纪的雕刻展示了荷兰捕鲸者在北极捕猎弓头鲸的场景。这座山似乎是挪威扬马延岛的贝伦火山,17世纪有大量的荷兰捕鲸船在此活动。

图22 1980年,荷兰考古探险队被允许在斯瓦尔巴群岛西北部挖掘荷兰捕鲸者的坟墓,从而披露了大量斯瓦尔巴群岛捕鲸期的新资料。坟墓获准挖掘后,出土了一些样本,之后就恢复了原状,现在看不到坟墓被挖掘的痕迹。

图23　1827年，挪威地质学家巴尔塔扎尔·马蒂亚斯·凯尔豪（Balthazar Mathias Keilhau）访问斯瓦尔巴时，绘制了一幅位于斯瓦尔巴东部的波默尔人狩猎站的画作。画中有一座木屋和东正教大十字架。大十字架既是宗教象征，也可以作为导航标志，方便从海上找到狩猎站。一只北极熊正在嗅挂在架子上的熊皮。画面背景中的山脉与今日当地的山脉很相似。

北极熊的毛皮。第一批捕猎者是来自俄罗斯西北部白海沿岸地区的波默尔人（俄语中意为"海边定居者"。）他们用自己带来的材料建造了小木屋和狩猎站，有时也会用些岸边的浮木。他们还竖起了东正教大十字架来显示其基督教信仰。

如今，几乎所有的十字架都已消失不见，但许多岛屿上仍然可以看到狩猎站和小屋地基遗址。

图24　斯瓦尔巴仅存的几个俄罗斯东正教十字架之一，其中底下的斜杠已遗失。

这些狩猎站的共同特点是角落里堆放着摞在一起的原木和红砖。砖是用来制造烤炉给木屋供热的。在斯瓦尔巴还可以看到这一时期的坟墓。直到19世纪末，越冬猎人主要来自挪威。他

图 25 斯瓦尔巴东部的波默尔人狩猎站遗址。底部的交叉原木显示出木屋的大小。破碎的红砖来自原来主屋角落的炉灶。

们在斯瓦尔巴周围建造了很多简易木屋。然后在接下来的数个冬季，为了得到珍稀的北极熊和北极狐皮毛，猎人们开始大肆捕猎。1973年签订的《北极熊保护协定》[9]规定，除格陵兰、北美和西伯利亚的原住民的传统狩猎外，禁止一切北极地区的北极熊狩猎活

图 26
a. 典型的猎狐陷阱，用岩石加重的木框由三块木头平衡，其中一块挂着诱饵，垫在木框下。当狐狸把头伸进木框咬住诱饵时，小块木头的移动使整个木框落下来把狐狸砸死，而不会伤害毛皮。毛皮才是猎人的主要目标。
b. 捕猎北极熊的"自杀式"陷阱。盒内的枪指向开口，扳机连到诱饵上。熊将头伸向开口，咬住诱饵，扳机被触发，熊头部中枪当场毙命。1973年之后，斯瓦尔巴的北极熊得到全方位保护。

动。如今，这些简陋的木屋大都遗留了下来，并且在斯瓦尔巴总督文化遗产办公室的保护下状况良好。这些木屋虽然不供游客使用，但我们可以通过木屋的简陋条件看出，就算到20世纪70年代，捕猎者对于在北极过冬的生存要求也不高。当然，这一时期的坟墓也说明了一个事实：并非所有的捕猎者都能够在此安全度过冬天。有些人装备和供应不足，死于饥饿和坏血病，也有些人死于小船或陆地上的意外。木屋已成为斯瓦尔巴历史和文化遗产的标志性建筑。除了木屋之外，在苔原周围还可以看到遗留的猎狐陷阱和捕熊工具。

1900年左右，人们开始在斯瓦尔巴进行商业采矿，其中最主要的矿产是煤，此外还有或多或少的铁矿石、大理石、黄金、石棉、锌等。除了煤，大多数矿物因为处于北极高纬度地区而并不具有经济效益，原因是很难运输。经过勘探和早期开采之后，矿

图27　想来斯瓦尔巴采矿大赚一笔的开采者带来了机械、矿工宿舍、采矿工具和设备，然而当希望落空，所有的设备都被抛弃于此。图中的机器是由欧内斯特·曼斯菲尔德于20世纪初从英国带来的，用于开采一流的大理石，但是最终开采出来的大理石质量不佳，根本卖不出去。今天机器仍矗立在被长期废弃的伦敦（或称"新伦敦"）遗址上。

工留下的遗址比比皆是，特别是在斯匹次卑尔根的西海岸。作为所有权声明的木屋和兼并标志、或深或浅的进山隧道、用来运输煤炭或岩石的铁路轨道和小型货车以及各式各样的工具和建筑物，为我们认识这个时期的斯瓦尔巴提供了便利。同时，这些物品也让我们看到，当时有一批人愿意在此寄托希望，并付出努力，只为实现从一片冻土之上发家致富的梦想，然而这通常只是黄粱一梦。

这一时期，有一个满怀希望的采矿者成了斯瓦尔巴历史上的一个传奇。欧内斯特·曼斯菲尔德（Ernest Mansfield）是一名怀着掘金梦的英国人。20世纪前20年间，他在斯瓦尔巴——之前曾在新西兰、澳大利亚和加拿大的黄金产区——挖矿。虽然他的雄心壮志是掘金，但今天最为人所知的是他建在康斯峡湾北侧布洛姆斯特兰半岛的大理石采石场，采石场的对面是新奥勒松定居点。

据说此地大理石质量极佳，因此人们对在此开办采石场的期望持续了好几年，后来才发现大理石一旦从地下取出，离开冻土

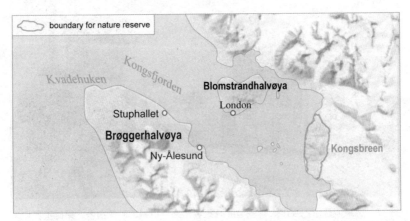

图28　图中地图展示了历史悠久的伦敦（London）大理石采石场和新奥勒松（Ny-Ålesund）科考站。这张现代地图反映了当前北极地区气候变暖的一个有趣特征："布洛姆斯特兰"（Blomstrandhalvøya）。这个名字的意思是"半岛"，但冰川（白色部分）向北方迅速融化，半岛实际上变成了岛屿。右下方绿线代表保护区边界线。

层后就会碎成碎片。然而曼斯菲尔德不失为一个具有传奇色彩的人物，围绕他的传说经久不衰。[10]

新奥勒松定居点位于康斯峡湾南部，由挪威矿业公司"国王湾"于1916年建立。煤矿开采初期进展顺利，但矿井1962年发生灾难性爆炸，导致开采被叫停。该定居点保留了下来，仍然由挪威政府管辖，现在是国际科学研究基地，来自10个国家（包括中国）的11个机构在此建立了科考站。每年夏季，此地就成了国际研究

图29 采矿者欧内斯特·曼斯菲尔德，他在斯瓦尔巴的伦敦（新伦敦）地区建了大理石采石场，初衷是寻找黄金，但发现了大理石，这似乎是一项利润可观的投资。不幸的是，大理石的质量欠佳，卖不出去。曼斯菲尔德因此成为斯瓦尔巴历史上的传奇。

图30 斯瓦尔巴群岛的新奥勒松定居点原本是一个采煤小镇。开采活动于1962年停止，后来挪威政府投入大量资源将其建成国际科学研究中心。

活动的热土。另外，新奥勒松拥有最多受法律保护的文化遗产建筑（28座，建于1946年之前）以及后来建造的具有高度遗产意义的建筑。因此，国王湾公司与挪威文化遗产部门密切合作，保护建筑物和定居点免受损害。

新奥勒松定居点位于北纬79°，是一个功能齐全的定居点。20世纪20年代，几支极具历史意义的北极探险队从此出发前往北极。美国人理查德·伯德（Richard Byrd）、挪威人罗尔德·阿蒙森和意大利人翁贝托·诺毕尔（Umberto Nobile）都是从此地飞往北极的，他们或是驾驶飞机［1925年，阿蒙森和美国的林肯·埃尔斯沃思（Lincoln Ellsworth）；1926年，伯德］，或是驾驶飞艇（1926年，阿蒙森、埃尔斯沃思和诺毕尔；1928年，诺毕尔）。除了作为北极探险的起点之外，如今新奥勒松吸引人的地方还有各种探

图31 新奥勒松定居点的飞艇系泊桅杆，1926年，挪威极地探险家罗尔德·阿蒙森驾驶"挪威号"飞艇飞越北极前往阿拉斯加，与其同行的还有意大利的翁贝托·诺毕尔和美国的林肯·埃尔斯沃思。1926年5月，"挪威号"上的16名男子成为第一批到达北极的人。飞行72小时后，"挪威号"降落在阿拉斯加的特勒市。

险的纪念，以及仍然矗立在苔原上的1926年的飞艇系泊桅杆。

虽然斯瓦尔巴位于北方，似乎不应受到第二次世界大战的影响，但事实上，有几次战役在该群岛留下了痕迹。挪威在朗伊尔城和斯韦阿格吕瓦的煤矿开采点以及俄罗斯（当时的苏联）的巴伦支堡都因遭到德国船只和潜艇的攻击而被摧毁。岛上仅存的痕迹包括朗伊尔城墓地里一名挪威士兵的坟墓，以及靠近城镇的两架德国飞机遗骸。其他的文化遗产离定居点较远，包括一架德国飞机，战时曾成功紧急降落在苔原上，之后被遗弃在此，此外还有或大或小的德国气象站遗址。对从北欧来的飞机和船舶来说，尽可能多地了解天气状况非常重要，因此，德国军队在斯瓦尔巴群岛以及格陵兰岛东部和法兰士·约瑟夫地群岛等北极西部地区都建立了人工和自动气象站。如今可以看到几个气象站遗址，其中

图32　德国"击剑"气象站于1944年9月在斯瓦尔巴东北部建成，一直运营至1945年9月。"击剑"是战时北极气象站的独特代表，但如果不改变建筑结构和原样，则很难对其进行维护。为了保护遗址免受人类影响，禁止游客靠近建筑物周围散落大量遗骸的区域。

保存最完好的位于斯瓦尔巴北部的东北地岛上。"击剑"气象站是一个受重点保护的文化遗产，尽管必须调整这两个木屋的原始结构才得以维持。这个地方很受游客欢迎，由于木屋比较脆弱，现在禁止游客靠近小屋和周围散落的遗骸，但是，从保护警戒线外可以很容易地观察该气象站。

斯瓦尔巴群岛最近被宣布为受保护的文化遗产是一个大型建筑群。为了庆祝 1957—1958 年国际地球物理年，在东北地岛的欣维克建了一个由 10 幢建筑组成的研究站。该研究站次年投入使用，此后只是偶尔用到。研究站属于挪威，2009 年，全站开始受法律保护。该建筑群对外开放，但是大部分建筑内部都是空的。

从朗伊尔城阿德文特达伦的矿场，空中运输系统将煤炭运往机场附近的霍泰尔尼斯特港口，这一路线在 2003 年已得到法律保

图33　1957—1958 年国际地球物理年，瑞典、芬兰和瑞士考察队在斯瓦尔巴东北地岛上建成的欣维克研究站。该站 1959 年后只是偶尔被使用。10 座建筑都属于挪威，建筑及其室内设备在 2009 年受到法律保护。其中一些建筑物上已经发霉，长出真菌，但斯瓦尔巴总督办公室在一定程度上对其进行了修复。

护。该运输系统一直使用到1987年，包括100个木架、用于支撑绵延10千米苔原的缆车以及其他建筑物，还有最重要的是朗伊尔城中连接新旧运输线的大型蜘蛛状中央索道站。

图34 位于朗伊尔城的中央索道站，煤矿中的煤先通过缆索从三个不同的方向输送出来，然后由一条缆索运到航运码头。图为1979年使用的运输系统。1987年，运输系统的最后一部分被关闭，缆索被拆除。支撑穿过苔原缆索的木制挂架依然矗立着，是受保护的历史遗迹。

由于斯瓦尔巴群岛拥有丰富的文化遗产以及壮丽的北极自然环境，挪威建议将群岛的大部分地区申报为联合国教科文组织世界遗产。从积极的方面看，申报世界遗产更能突出遗址的普遍价值，促进对其的保护。另一方面，众所周知，一旦被评定为联合国教科文组织世界遗产，参观这些遗迹或地区的游客人数将显著增加，而对于环境脆弱的北极高纬度地区而言，如今造访斯瓦尔巴群岛的游客似乎已经够多了。

南极洲的文化遗产

南极洲在历史上没有原住民,也没有任何矿产勘探和开采活动,但和北极地区有着一段类似不过短得多的历史,那就是曾有捕鲸者、海豹捕猎者、探险家以及科学家的足迹。南极半岛的海豹捕猎活动始于 19 世纪早期,很快便出现了过度捕猎的问题。捕鲸活动随后于 20 世纪初兴起,在南极洲和亚南极洲岛屿上也建起了捕鲸站,后来同样导致过度捕猎。

图 35 南极洲欺骗岛捕鲸湾上 20 世纪初捕鲸站的遗迹。从 1906 年至 1931 年,鲸加工就在这里的海岸上进行,即利用鲸脂、鲸骨熬制鲸油,今天的遗迹是后来留下的。欺骗岛是一座火山岛,1969 年的火山喷发破坏或摧毁了捕鲸者的墓地、房屋以及其他建筑。2005 年,考虑到自然环境、文化遗产、研究和旅游业等因素,欺骗岛被列为南极条约体系下的南极特别管理区。

南乔治亚

南乔治亚岛是英国在南大西洋的海外领土[11],地处南纬 54°—55°。该岛位于南极地区(南纬 60° 以南)北部,但由于它是 20 世纪上半叶南极和南大西洋捕鲸活动最重要的基地,因此被包括在南极特别管理区内。早在 1786 年,海豹捕猎者就在这里的避风港作业了。1904 年,捕猎者在古利德维肯湾建立了第一个捕鲸站。古利德维肯湾得名于挪威语中的"gryte"一词,意指熬制海豹油的罐子,在这里就发现了此种罐子。有四个捕鲸站成了大型工业园区,不仅处理每一头被拖上岸的巨鲸,还为那里的工作人员提供各方面的服务。除了生活营房外,还有配套的教堂、电影院、图书馆、体育设施、食堂以及墓地。在 20 世纪 60 年代中期捕鲸活动结束后,这些捕鲸站便被废弃了。由于自然条件恶劣,以及沿途船只的洗劫,这些捕鲸站遭到严重破坏。如今,为了防止遗迹受到进一步破坏,它们已被认定为需要得到保护的捕鲸时期重要工业遗迹。遗憾的是,由于破败的建筑物中有大量石棉,还面临着倒塌的风险,对游客来说太过危险。不过,这些遗迹已经通过

图 36 南乔治亚的利斯港捕鲸站,运营时间为 1909—1965 年。大量的工业建筑和其他建筑物一直受到自然环境退化和游客破坏的影响。南乔治亚政府现已制定了该岛和捕鲸站的文化遗产战略,旨在防止人类影响造成进一步破坏。该站周围有 200 米的保护禁区,防止游客进入有石棉和坍塌建筑物的危险区域。利斯港捕鲸站和岛上其他历史悠久的捕鲸站已经通过 3D 扫描和摄影进行了全面记录。

3D 扫描被完全记录了下来，游客可以"虚拟"参观这些捕鲸站（参见 https://www.youtube.com/results?search_query=geometria+ltd），也可以从小船上眺望，或者在 200 米禁区的外围步行参观。在亚南极洲地区的原始自然景观中看到如此大量的工业废墟可能会让一些人感到不快，但这些遗迹实际上占地十分有限，与雪山和草丛茂密的海岸壮丽景色无法相比。这里是海豹和企鹅的世界，也是重要的历史信息来源地，既可以帮助人们了解捕鲸方法，也有助于了解该行业如何在离家万里的地方建立这样的工作场所，还能让游客了解几十年前在与世隔绝的亚南极洲岛屿上从事捕鲸活动是什么体验。

南极洲历史上最著名的阶段要数 1897—1922 年间探险的"英雄时代"了。在这一时期，许多人将他们的胜利或悲剧载入了世界史中，例如，罗尔德·阿蒙森（挪威人，在 1911 年成为抵达南极点的第一人）、法比安·戈特利布·冯·别林斯高晋（Fabian Gottlieb von Bellingshausen，俄罗斯人，1819—1821 年完成环南极大陆的航行）、卡斯滕·博克格雷温克（Carsten Borchgrevink，挪威人/英国人，1899 年成为在南极大陆越冬第一人）、让-巴蒂斯特·沙尔科（Jean-Baptiste Charcot，法国人，1903—1910 年间领导了两次南极洲探险）、阿德里安·德·热尔拉什（Adrien de Gerlache，比利时人，1898—1899 年间领导了首支南极洲越冬航船探险队）、罗伯特·福尔肯·斯科特（Robert Falcon Scott，英国人，1912 年领导了到达南极点的第二支探险队）以及欧内斯特·沙克尔顿（Ernest Shackleton，英国人，1901—1922 年间开展了几次南极探险）等。

在资源开发者和探险家之后接踵而至或相伴而来的是科学家，科学家们主导了如今的南极活动。1957—1958 年的国际地球物理年和由此产生的1959年《南极条约》将南极洲定义为致力于和平与科学的大陆。南极洲现有来自全球各大陆29个国家的70个永久科考站。随着旅游业扩张到全球几乎每个角落，南极洲的旅游业也在迅速增长和扩张。大多数游客选择乘船前往南极，而南极半岛是最受科学家和游客欢迎的地区。虽然由于冰况复杂、条件严峻而难以到达，但罗斯海地区及其具有历史意义的小屋对游客来说很有吸引力。目前，游客还可选择乘飞机前往南极。1977—1979年期间从新西兰和澳大利亚有飞往南极的观光飞机，但新西兰航空的一架飞机于1979年11月在埃里伯斯山坠毁，造成257人丧生，随后这些航班被取消。1994年，澳洲航空恢复了从澳大利亚前往南极的包机，目前一些小公司有航班将人数有限的探险（体育）团队从南美洲送往南极洲，进行滑雪和登山活动，但不对普通游客开放。科学家们在夏季飞往几个较大的基地，包括南极点的美国阿蒙森－斯科特基地。预计未来飞往南极大陆的旅游航班将继续增加。

一些标志性的极地探险

▼

除了1910—1912年的日本南极探险之外，历史上亚洲的极地活动非常少。当时，日本探险队由日本陆军中尉白濑矗（Nobu Shirase）率领，乘坐"开南丸号"前往南极。尽管相对而言，他们对在南极洲遇到的状况并未做好充分准备，但还是成功抵达了罗斯海，并碰巧遇到了由罗尔德·阿蒙森率领的挪威探险队，后者刚刚成为首支抵达南极点的探险队，处于归途当中。由7名日本人组成的所谓"快速巡逻队"在罗斯冰架（南极冰架）登陆，然后向南走到南纬80°05′，同时另一组人前往爱德华七世半岛海岸的亚历山德拉皇后山脉的缓坡。之后，"开南丸号"回到了日本。尽管队员在行程的最南点留下了纪念信息，但这些信息很快便在风雪中销声匿迹，难觅行踪。本次探险活动也没有留下任何文化遗产。日本在本州西北部秋田县修建了白濑南极探险纪念博物馆纪念本次探险。

南极大陆的首批建筑物

19世纪早期的海豹捕猎者在南极半岛地区建造了简易的住所，其遗址至今仍可以看到。南极大陆建立的第一批坚固的建筑物今天仍然矗立着——这在任何其他大陆都是极为罕见的。挪威人卡斯滕·博克格雷温克1898—1900年率领的一支英国探险队成为首支在南极大陆过冬的探险队。该探险队留下的遗址位于罗斯海西入口处的阿达尔角，有两间预制木屋，由位于挪威奥斯陆郊外的一家木材公司建造。当探险队乘"南十字号"离开时，这些木屋就留了下来,直至今日仍是重要的文化遗产。新西兰遗产组织南极遗产信托基金致力于保护木屋及1900年遗留在内的物品,但由于其偏远的地理位置、极端的气候条件以及后勤问题,这着实是一个艰巨的挑战,尤其是必须尽量避免干扰目前生活在阿达尔角遗址大部分地区的大量筑巢企鹅,甚至包括那些沿着木屋墙壁和在其中一间开放式木屋内筑巢的企鹅。依靠挪威和英国等国的额外资源,有一个项目已经在阿达尔角开始进行,该项目将在未来许多年内保护这个独特的遗产地。在2017—2018年南极的夏季期间,遗址修复队得到了中国第34次南极科学考察队的支持,后者派遣了40人及其Ka-32重型直升机与破冰船"雪龙号"一起帮助解决后勤问题。

图37　这两座位于南极洲阿达尔角的木屋是南极洲主大陆上的首批建筑,由挪威人卡斯滕·博克格雷温克率领的英国科考探险队于1899年建立,木屋是在挪威预制的。这支探险队是首支在南极大陆过冬的探险队。木屋及屋内的物品目前由新西兰南极遗产信托基金保存,挪威和英国提供了相应资源,但阿达尔角交通极其不便,并且只能作短暂停留。

图38 中国破冰船"雪龙号"。

阿达尔角还有另外两项重要的文化遗产，即南极大陆上已知的第一座坟墓和罗伯特·斯科特的北方小队居住的一间小屋的遗址。后者是斯科特团队在1911—1912年夏季为了科研工作从罗斯岛的主要基地转移过来的。坟墓是博克格雷温克领导的探险小队中动物学家尼科莱·汉森（Nicolai Hanson）的安息之地，他在南极大陆越冬期间因病去世。

法兰士·约瑟夫地群岛的一个冬天

极地历史充满了关于生存和创造力的奇妙故事，至今仍然激励着我们。当这些故事可以直接与某一地点、遗址、船只或住所的遗迹产生联系，当站在这些遗址旁，任凭想象力回溯数年或数百年时，现实能够产生深远的影响，并有助于我们珍惜这些故事和物质遗产。其中一处遗址位于俄属北极高纬度地区法兰士·约瑟夫地群岛的中部。

挪威人弗里乔夫·南森（Fridtjof Nansen）于1893—1896年率领探险队乘坐特别设计的船只"弗雷姆号"穿越了北冰洋甚至北极点。彼时的专家认为南森的计划非常危险，甚至无异于自杀，因为当时已有许多木船受困北极海冰并受到挤压的事例，刻意让船被海冰冻住的计划被视为疯狂之举。南森是想证明北冰洋之间存在一股东西走向的洋流——他做到了——并同时调查了北极点周边地区是否存在陆地，当时人们对此一无所知。在"弗雷姆号"航行期间测量到的令人惊讶的海洋深度——最深近3 000米——足以证明该地区没有陆地。"弗雷姆号"独特的设计和造船用的巨大

图39 挪威奥斯陆博物馆展出的极地船"弗雷姆号"。该船是为挪威科学家弗里乔夫·南森1893—1896年的探险专门设计和建造的，探险队准备从白令海峡西部横穿北冰洋到达挪威北部以证明存在一股东西走向的洋流横穿北冰洋，并开展许多其他的观测和测量活动。"弗雷姆号"之后于1898—1902年被奥托·斯韦德鲁普（Otto Sverdrup）率领的探险队用于探索加拿大东北部地区地图上未标注的岛屿，又于1910—1914年用于罗尔德·阿蒙森的南极探险。该探险队乘坐狗拉雪橇首次到达南极点。挪威奥斯陆博物馆专门为"弗雷姆号"建造，由挪威国王在1936年宣布对外开放。博物馆还拥有一艘较小的船"约阿号"，在1903—1906年罗尔德·阿蒙森率领的一次探险行动中，它成了第一艘驶过西北航道的船。

木材使她能够承受海冰的巨大压力。今天，人们能够在奥斯陆的弗雷姆博物馆里里外外仔细地观察这艘船。

在对自己的诸多理论进行验证后一年，南森再也按捺不住内心的躁动，便与亚尔马·约翰森（Hjalmar Johansen）一起离开了"弗雷姆号"，在狗拉雪橇的帮助下向北极点进发。他们不得不在北纬 86°14′ 处返回，向南穿过不断变化的不均匀冰层到达陆地，当时还不知道这是法兰士·约瑟夫地群岛。他们与其说建造了一间小屋，不如说是在地上挖了个洞，然后在这里度过了 1895—1896 年的冬天。他们用海象的肩胛骨和滑雪板的末梢在坚硬的地面上铲出了一个长 2 米、宽 2 米、深 1 米的坑，然后筑起了 1 米高的石墙，并用一根浮木支撑起几张海象皮作为屋顶。两人在

图40 弗里乔夫·南森和亚尔马·约翰森在法兰士·约瑟夫地群岛上过冬时居住的简易"小屋"遗址，他们在苔原上铲出一个 2 米 ×2 米的坑，筑起 1 米高的石墙，用一根浮木支撑起几张海象皮作为屋顶。他们的故事讲述了如何在设备极其稀缺的情况下在北极冬季存活下来，令人惊叹。现在游客能在游轮上参观该遗址，但遗址周边脆弱的苔原植被很容易受到人类的影响，因此应严格控制当地旅游业规模。

寒冷和黑暗中躺了8个月，共用一个破旧的驯鹿皮睡袋取暖，靠吃海象和北极熊的肉和脂肪维生。在类似条件下，其他在极地过冬的人经常死于饥饿和坏血病，而南森和约翰森始终保持神智清醒和身体健康——实际上他们的体重还增加了约10千克！春天一到，他们继续向南穿过群岛，直到幸运地遇见一支由弗雷德里克·杰克逊（Frederick Jackson）领导的英国探险队，并乘坐后者的船返回挪威。回到挪威仅仅一周后，"弗雷姆号"及其船员在航行了3年后也回到了家乡。

除了奥斯陆的"弗雷姆号"之外，法兰士·约瑟夫地群岛越冬地点的遗迹至今也依然存在，现在仍可以看到地面上的一个坑，周围是石头和浮木。正如南极洲的简易小屋与周围环境完全融合在了一起，南森越冬地点的遗迹也已与周围的自然环境融为一体。小屋的遗迹和周围的自然环境都需要保护，以防游客过多对遗址造成破坏并加速其状况恶化。

巴伦支海的名称由来

荷兰航海家和制图师威廉·巴伦支（Willem Barentsz，英文通常写成 Barents）于1594年、1595年和1596年向北方和东方航行，寻找传说中前往中国的东北航道，每一次都被俄罗斯西北海岸的海冰阻挡。在1596年的最后一次航行中，他首先到达了一座岛屿，探险队将其命名为熊岛（斯瓦尔巴群岛最南端），然后向北航行到斯瓦尔巴群岛斯匹次卑尔根岛的西北角。接着，巴伦支返回熊岛并向东航行到新地岛，由于海水冻结和冬天的黑暗到来，他和船员被迫在那里过冬。16名船员用从船身拆下的木料建造了一个7.8米×5.5米的木屋，称之为保命屋。巴伦支于1597年6月去世，当时船员们正乘着两艘小船向南行进。最终有12名船员幸存下来，并被带回荷兰。新地岛上的木屋遗址至今尚存，依然清

图41　新地岛上荷兰人的越冬住所遗址——"保命屋"——由威廉·巴伦支及其船员为渡过1596年与1597年之交的冬天建造。

晰可见。但遗憾的是，由于几个世纪以来自然因素、纪念品收藏者以及考古学家的原因，这些遗址均遭到严重破坏。尽管如此，它仍是一个重要的遗产地，应该受到保护，防止人类行为对其造成进一步的破坏。

威廉·巴伦支探险和殒命的故事通过一系列以其名字命名的地区为我们留下了额外的非物质文化遗产，包括巴伦支海、斯瓦尔巴群岛巴伦支堡的俄罗斯人矿区以及挪威北部和俄罗斯西北部的巴伦支地区。

图42　俄罗斯斯瓦尔巴群岛巴伦支堡的采煤定居点。根据《斯匹次卑尔根条约》(斯瓦尔巴群岛)，签署该条约的任何国家的任何公民在斯瓦尔巴群岛的经济活动中都拥有与挪威人相同的权利。俄罗斯(以前是苏联)是除挪威之外唯一一个在该群岛有定居点的国家。该定居点使用煤炭供能，在雪白的背景下可以清楚地看到煤烟污染。

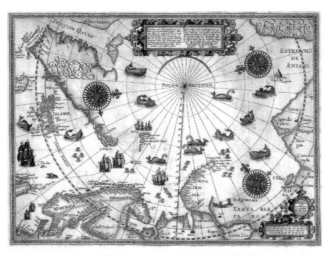

图43　威廉·巴伦支第三次航行的地图，航行中发现了斯瓦尔巴群岛的一部分并绘制了地图，航行最终以被迫在新地岛越冬告终。

一次探险、三处遗址和一艘沉船

1901年至1904年，瑞典地质学家尼尔斯·奥托·古斯塔夫·努登舍尔德（Nils Otto Gustaf Nordenskjöld）率领的瑞典南极探险队乘着"南极号"对南极开展探险活动，由挪威人卡尔·安东·拉森（Carl Anton Larsen）担任船长。探险队的目的是进行地理探索，努登舍尔德和5名船员按照计划在南极越冬，他们在南极半岛附近的小岛雪丘岛上建造了一间木板屋。"南极号"和拉森在冬天过后回到雪丘岛接这些船员时遇到巨大的困难，海冰令他们无法通行。3名船员留在希望湾，并为在雪丘岛的船员提供了一个物资点，以防"南极号"无法抵达雪丘岛。这艘船继续与海冰斗争，最终却被困其中，船员不得不将船遗弃。这并非唯一一艘受到北极和南极极地海冰挤压并沉没的船。船长拉森带着他的船员安全到达普雷特岛，在那里建起一间临时石屋。一名船员丧命于此，被埋在附近。留在希望湾的3名船员被迫自己建立了紧急冬季避难所，因为"南极号"再也没有回去接他们。3个不同的地方

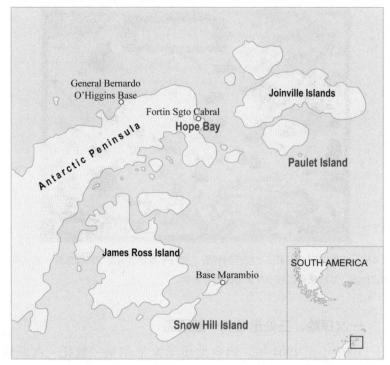

图44　瑞典南极探险队1901—1904年之间的三个越冬地点：雪丘岛（Snow Hill Island）、希望湾（Hope Bay）和普雷特岛（Paulet Island）。

图45　两处遗址的照片

a. 普雷特岛上的石屋遗址，拉森和其他19名船员在他们的船受到南极海冰挤压后，在此度过了1903年的冬天。

b. 雪丘岛上的石屋遗址。这座瑞典探险队的房屋于1902年建在一个冰碛小丘上，小丘由早期的冰川活动留下的沉积物构成。然而，这个小丘并不稳固，因此船员试图将小丘的边缘稳固下来。

有3组船员，都在等待冬天结束，希望能够获救。令人惊讶的是，他们最终都获救了，既是靠运气，也有赖于由伊里萨尔（Irizar）中尉领导、被派去寻找他们的阿根廷轻巡洋舰"乌拉圭号"。

仅仅这一次探险就留下了雪丘岛上的木屋、普雷特岛上的石屋遗址、船员奥勒·温内尔斯加德（Ole Wennersgaard）的坟墓以及希望湾紧急小屋的废墟，所有这些遗址都被列为南极保护文化遗产。此外在海冰下还有"南极号"的残骸。

// 极地文化遗产

南极海豹猎人和早期因纽特人的遗址

▼

作者在此并非要暗示19世纪早期南极洲的海豹捕猎者可以与北极高纬度地区的早期因纽特人相提并论，但两者住所遗址的确有相似之处，也都丰富了今天我们对极地文化遗产的理解。当然，北极地区的生活方式可能因地区而异，但在某些比较中存在相关性。不论是在南极的海豹捕猎区还是在早期因纽特人的群体中，修建房屋都并非为了长住。在北极高纬度地区，早期因纽特人根据狩猎和捕鱼的食物供应以小型家庭团体为单位进行迁移。如果采用定居的生活方式，附近的猎物狩猎完后他们便会面临挨饿的困境。因此，房屋的修建非常简便，可以供财产很少的流动家庭或小群体居住。在冬季，许多地区可以用雪块（雪屋）建造相对温暖而坚固的房屋，而在夏季，帐篷或其他普通的住所就足够了。人们在修建住房时不会将材料从一个地方搬运到另一个地方，而是尽可能地因地制宜。在草木不生的北极地区，石头、土壤、动物骨骼和毛皮均能作为建筑材料，尽量做到就地取材。而且住所的大小不会超过家庭所需的面积。这也是一种控制取暖的方法：当室内温度为几摄氏度的正常水平时，一盏简易的鲸脂灯和人或狗的热量足以取暖。当一个家庭迁走时，一堆固定毛皮帐篷的环状大石或房屋中央作为壁炉的一堆石头会被留在原本空空如也的苔原上。今天，如果看到这些遗迹，首先会让我们思考这处遗迹的历史到底有多悠久——石头不会像木头那样老化，上面也几乎不会有植被生长——但专业知识有助于揭示石头布局背后的意义，并深入了解这种适应极端气候的文化。

19世纪初期，为了捕猎海豹、剥制毛皮以便运回贩售，在南极半岛地区活动的海豹猎人有时需要在寒风凛冽、冰天雪地的海岸度过数天或数周的时间。若遇上恶劣天气，他们有时甚至需要在此熬过数月乃至整个冬季。与前文所述的因纽特人的住所一样，他们在搭建住所时也是就地取材，无论是岩壁上的洞穴还是用石头、兽骨打磨制成的石材，只要合用，都会被他们用来搭建栖身之所。幸运的话，有时还能从船上找到些闲置的木料或者船帆。在过去的150—200年间，可以用来证明这里曾经有人居住的工具、器具等物品大多已像其他因纽特人遗迹一样随时间消逝，这就给我们理解和解读这些遗迹造成了一定困难，这些遗址的布局和材料虽然平平无奇，但对于我们理解南极历史极为重要。

图46
a. 19世纪南极海豹捕猎区域。
b. 加拿大北极地区的一个早期因纽特人聚居区。

无论北极还是南极，这种区域都没有得到足够的保护，因为它们太不起眼，所以人们对其重要性和起源都没有足够的认识。现代露营者为了方便搭建帐篷，有时会移动北极地区的石头遗迹，而南极遗留的卵石堆遗迹则可能会被研究当地地质历史的地质学家移动或者分开。在保护极地文化遗产这一最不起眼的部分时，信息是更好地理解其价值的关键，哪怕只是不寻常的石块布局也值得我们注意。

包括上述遗址在内的几乎所有北极高纬度地区和南极洲的遗址都有一个共同点：考古发掘经常可以在地表下发现更多的证据和信息，但传统考古意义上各种不同文明在不同历史时期堆叠形成的"文化层堆积"则较为鲜见。这一方面是因为永冻土的存在使得当地人很难挖掘较深的地基，甚至连坟墓都难以挖深，另一方面是因为在历史上，对于这些地区的开发程度较低，也就没有必要像其他古老城镇那样，在千百年间不断开发，在旧城的基础上建设新城。"所看即所得"可以适用于类似图46、47的所有地点。但即便如此，通过发掘表面较浅的区域也能给我们带来惊喜。

图 47

a. 发掘前的南极象角海豹捕猎区域。

b. 挖掘后的现场，可以看出即使是这里也存在较浅的文化层。

早期文化和活动的物质证据

▼

通过上述段落的介绍可以看到极地地区拥有丰富的历史。北极地区的历史绵延数千年，留下了丰富的文化遗产，它们成为这里发生过的种种故事的见证。虽然在极地壮丽的自然环境面前，这些文化遗产似乎不值一提，但它们对人类历史完整性的重要价值丝毫不亚于北非、南美的金字塔或者中国的万里长城。如果没有它们，我们对人类从亚洲到阿拉斯加北部、再到加拿大以及格陵兰的迁徙之路的了解将大为减少。我们将难以了解最早期人类历史的完整情况，在人类的历史长河中，他们在艰苦卓绝的环境中登场，又由于环境恶化无法生存而退出历史舞台。我们也将更难想象，在这样一个让数百名探险家殒命的环境中，早期人类先驱是如何生存下来的[12]。

北极

如前文所述，北极地区的文化遗产大致可以分为两类：原住民遗产和源自南方文化的遗产，后者通常来自个人或较小的群体，他们北上的目的主要是通过狩猎、捕猎、捕鱼、捕鲸及采矿等方式来获取自然资源，其他目的还包括勘探、科研及社会工作——甚至还包括战争和"冷战"。北极地区多样化的文化遗址和景观对人类具有重要价值，无论是在个人层面还是国际层面上都是如此。它们是我们了解人类在各个历史阶段如何与北极自然环境互动的知识宝库。它们反映了人类踏足北极的动机，以及人类对北极的认识和解读。它们是那些描述努力奋斗、获取成功的故事的灵感源泉。原住民同时受到当地非物质文化遗产和现代生活的影响，两方面的因素在历史的进程中共同构成了其自我观念和地方认同的基础，并将对未来继续产生影响[13]。

在国际舞台上，北极地区与许多其他具有重要国际影响力的地区相比颇为不同。北极地区鲜有壮观雄伟的建筑，其文化属性与自然属性也没有那么明显的区分。与此同时，该地区的早期遗迹未因农业的发展或后期文化层的出现而被掩盖在历史的尘埃之中，当地的气候条件也让其有机物质得到了较好保存，这在南方地区难以看到。此外，代表北极早期探索的遗址常常被神化，成为历代文艺作品中的元素[14]。

图48　英国油画《西北航道》，由米莱斯（J. E. Millais）于1878年创作。这幅画描绘了人们想象中英国尝试开辟西北航道的那段历史，极具象征意义。画中描绘了老水手向女儿讲述西北航道故事的场景，背景中则可以看到加拿大北部海岸航海图、英国海军旗帜以及远征队日志，在画中旗帜的后面挂着两幅油画，一幅描绘了舰船受困海冰的场景，另一幅则是英国海军英雄纳尔逊勋爵（Lord Nelson）的肖像。

北极高纬度地区归属于五个国家：俄罗斯、挪威、丹麦王国（格陵兰）、加拿大和美国。各国在文化遗产方面的法律和政策不尽相同，这就给我们全面了解文化遗址造成了极大挑战。文化遗址因为有固定的位置，理应比四处移动的北极熊或海鸟群落更易统计，但因为各国在统计方法、定义以及获取信息的方式等方面存在差异，使统计面临着相当大的困难。《斯瓦尔巴群岛环境法》[15]规定，以1946年1月1日为界，在此日期前出现在北极圈内的所有固定和移动文化遗产，无论其来源及保存情况，均自动受到法律保护。这就导致第二次世界大战或1946年之前的国际科学活动留下的垃圾也会得到与17世纪早期捕鲸站或19世纪初捕猎者越冬小屋遗迹同等的保护。由于其他国家对于文化遗产法律保护的适用范围有不同意

图49 斯瓦尔巴群岛遗址：a.狩猎舱；b.采矿机械；c.炼油炉（从鲸脂中炼制鲸油）；d.鲍末狩猎站遗址

见，这种一刀切的措施给泛北极文化遗产定义产生分歧埋下了隐患。斯瓦尔巴群岛共有2684处受法律保护的遗址和古迹被收录进了挪威全国文化遗产保护数据库[16]。其中包括前文提到的两个历史较短的遗址：一是二十世纪五六十年代建设的大型煤炭运输系统，该系统连接煤矿与货运码头；二是1957—1958年国际地球物理年建设的科学站，包括10幢独立建筑。这也引发了新的问题，即包含多处古迹的遗址应当按一处还是多处登记。举个例子，假设某处17世纪建成的捕鲸者墓地包含20座坟墓，在登记时被登记为一处遗址，那么随着海岸侵蚀，部分坟墓被冲入海中，墓地规模缩小，该如何登记呢？如果不单独登记每座坟墓，就很难量化实际损失。

图50　格陵兰东北部的捕猎者小屋（a）与同一时期斯瓦尔巴群岛的捕猎者小屋（b）非常相似，二十世纪二三十年代建成的小屋尤其如此。

而《格陵兰文化遗产法》[17]以1900年为界，将此前的文化遗产自动纳入保护范围，不包括20世纪20年代到20世纪40年代的丹麦和挪威的捕猎者小屋，这些小屋是斯瓦尔巴群岛保护区的一大特色遗产。上述两个区域的小屋由同一类人在同一时期建成，设计及材料也颇为相似。令人高兴的是，格陵兰岛东北部的小屋并未被人遗忘，近年来，在文化遗产主管部门的许可下，私营企业对这些小屋进行了较大程度的复建。

北极许多地区的原住民遗产通常规模较大,不仅包括长期或短期的居住遗址,还包括狩猎区域、墓地及宗教场所。这些遗址的历史可以追溯到公元前数千年。例如加拿大育空地区和西北地区的高山冰川地区,这里留有的驯鹿狩猎证据通过放射性碳元素断代,可以追溯到9 000多年前[18],为驯鹿狩猎设置的因努伊特石堆和帐篷遗迹可以追溯到4 000多年前,遍布阿拉斯加阿基亚科湖区的大片区域[19]。格陵兰北部考古遗址中发现的"独立 I 古爱斯基摩文化"证据可以追溯到4 200—4 000年前[20]。

图51 独立 I 古爱斯基摩文化遗址。这些遗址在格陵兰岛最北部的裸露石质地貌上并不容易发现,但在陆地区域的中部发现了两处居住地遗址。

南极

如前所述,南极半岛最早期的海豹捕猎遗址在很多方面都与北极的古爱斯基摩遗址具有相似之处。这些遗址虽然仅能追溯到19世纪早期,但仍能提供一些证据帮助我们理解当时的人们是如何在南极恶劣的夏天工作、生存,并通过捕猎海豹、制作海豹皮谋生的。与北极地区的古爱斯基摩遗址一样,南极最早期的遗址虽然包含一部完整的人类活动史的关键信息,但其本身往往很不起眼。在南极文化保护的另一方面,"英雄"探险家和早期科学家的木屋与屋中的双层床、罐头食品、雪橇犬犬绳、科学设备和其他物品一起得到了保存,让游客及电影、照片的观众能够了解这些无畏的勇士当时的生活和工作环境。对于他们而言,在探求终极地理和科学目标的旅途中,冻掉手指或脚趾的事故时有发生,有时甚至还要冒着失去生命的风险。

木屋的意义远大于其本身

▼

在许多人眼中,南极历史主要就是所谓"英雄时代"探索和科学考察的历史。英雄时代通常指1897年阿德里安·德·热尔拉什·德·戈梅里（Adrien de Gerlache de Gomery）率领比利时南极探险队乘坐"比利时号"到访南极,至1922年欧内斯特·沙克尔顿殒命南乔治亚岛之间的这段时期,其间充斥着各种英雄人物的探险故事,为数以百计的著作提供了素材,见证了坚持、进取、痛苦、死亡、胜利、或好或坏的计划以及让我们读之手不释卷、听之津津有味的幸运或不幸的故事。其中一些人物除了留下诸多可歌可泣的故事,还在南极大陆留下了有形的物质遗产。最具代表性和知名度的物质遗产非"具有历史意义的小屋"莫属,它们具有后期南极半岛地区科研基地小屋难以比拟的崇高地位。

南极大陆现存历史最悠久的小屋包括:卡斯滕·博克格雷温克在阿达尔角建造的小屋（1899—1900）,罗伯特·斯科特率领的探险队在哈特角建造的探险小屋（1901—1904）,欧内斯特·沙克尔顿在罗得斯角建造的"猎人号"探险小屋（1907—1909）,斯科特在埃文斯角建造的特拉诺瓦小屋（1910—1913）,尼尔斯·奥托·努登舍尔德在雪丘岛建造的小屋（1902—1904）,以及道格拉斯·莫森（Douglas Mawson）率领的澳大利亚探险队在丹尼森角搭建的探险小屋（1912—1913）。上述小屋均已被认定为重要的南极世界遗产,从而尽可能地减少时间流逝、游人增多等因素对小屋及屋中物品的影响。新西兰南极遗产信托基金与谷歌合作[21],将斯科特在埃文斯角的特拉诺瓦小屋通过网络平台向公众开放,从而让更多人能够通过虚拟游览的方式体验这些宝贵的遗产。许多博物馆、艺术画廊和重要的遗产网站也在使用这种方法，在让更多人有机会体验景点风貌的同时，避免给景点造成额外的压力，这样一来，那些原本永远不可能有机会亲身到访这些景点的人也能获得身临其境般的体验。虽然这些小屋及其中物品本身并无太多惊艳之处，但由于地处南极，加之在其中发生过的历史事件，它们也就成了特别具有标志性意义的国际遗产。

本文所称文物（或人工制品）是指具有历史意义的可移动物品，其本身应当具有一定年份，或与某些重大历史事件或重要历史人物有关。在考古发掘过程中发现的工具、器具、衣服或部件均称为文物。类似地，与较近期发生的具有历史意义的事件或人物相关的物品也是文物而不仅仅是普通物品。在历史悠久的南极

图52　斯科特的埃文斯角小屋内景。从图片可以看出，在这种具有重大历史意义的小屋中有大量历史文物需要得到更专业的保护，从而避免遭受更多损失。以食品罐为例，由于其包含纸质标签、锡罐和罐内物品等三个部分，对其进行保护时就需要采用三种不同方式。

小屋内及其附近区域，常能发现大量探险队回家路上遗留下的文物，包括食品罐子、瓶子、化学品、科学仪器、书籍纸张、家具、雪橇犬犬绳、包装箱、皮革衣物和厨房用具，以及在南极干冷的气候中被冻成木乃伊的狗和企鹅的尸体。不幸的是，多年来，许多文物没有逃过被游客或其他人移动、摆弄的命运，通过对比历年照片可以清晰发现这一现象。目前，除非出于保护目的，以任何其他目的破坏或取走小屋中的文物都将被认定为违法行为。近年来，新西兰与英国的南极遗产信托基金已展开行动，对相关文物展开大规模全面保护。以斯科特的埃文斯角小屋为例，新西兰南极遗产信托基金一项为期七年的文物保护计划在 2015 年宣告完成，通过该计划成功对 11 500 多件文物实施了保护[22]。

《南极历史遗址及古迹名录》

如上所述，南极属于国际区域，不受任何单一国家管辖，也不归属于任何单一国家。南极治理基于南极条约体系实现，在所有缔约国达成共识的基础上协同行动。该条约于1959年制定，当时共有12个缔约国。目前，已有53个国家签署或加入该条约体系，并每年召开会议讨论相关议题。但在上述国家中，目前仅有29个国家具有协商国地位或投票权，某国是否具有协商国地位取决于该国是否在南极开展实质性的科考活动，与其是否对南极提出领土主张无关。1968年，南极条约协商会议提议汇编南极的历史遗址及古迹名录，为南极条约区域内大陆及岛屿上的具有重大意义的物质文化遗产编列清单。该历史遗址和古迹名录现已收录92处古迹，但必须要指出的是，如果从专业的文化遗产角度评判，如根据联合国教科文组织世界遗产系统的标准评判，名录中多处古迹和遗址或许有些名不符实。南极遗址应当对整个世界都具有重大意义，而不仅限于对某国具有纪念意义。近来，在体系内也开始讨论让专业文化遗产知识扮演更重要的角色，从而提升南极文化遗产的地位，使其能与自身得到的专业保护更加相称[23]。列表上的第87号历史遗迹是一块青铜纪念匾，它标明了第一座提供长期服务的德国南极科考站"格奥尔格·福尔斯特"的位置。该科考站位于毛德皇后地的施尔马赫绿洲。这块纪念匾保存完好并且固定于该区域最南部的石墙上。该科考站在1976—1996年投入使用，1996年2月12日拆除行动成功结束后，整个区域被彻底清理。虽然科考站曾经存在于此值得纪念，并且这也是人类南极历史中

图53 历史遗址和古迹名录第17号遗址，埃文斯角的十字架纪念碑，1916年落成，纪念在附近殒命的三名英国探险家。

图54 历史遗址和古迹名录第76号遗址，佩德罗·阿吉雷·塞尔达科考基地遗址，该站原为智利建设的气象和火山活动检测中心，位于南极钟摆湾欺骗岛，1967年和1969年毁于火山爆发。

的重要篇章，但有争议的是，最近才在此地竖立起来用以纪念科考站的纪念匾是否能够被视作值得保护的文化遗产。科考站的建立才是真正的"遗迹"。即使在纪念匾消失的情况下，这一段历史依旧存在，而纪念匾可以在不损失任何重大文化价值的情况下进行更新和替换。这仅是历史遗迹列表上遭到专家质疑的众多遗迹之一。

前文提到的具有重大历史意义的小屋已被收入历史遗址和古迹名录，而同样在南极人类历史中留下浓墨重彩的海豹捕猎区域则尚未被收入。理想情况下，南极遗产名录包括的纪念物和遗址等可以分为两类，一类是像联合国教科文组织世界遗产名录包括的那种具有重大国际意义的遗产，另一类是具有重大国内意义的遗产。近年还有观点认为，如果将南极的一些遗产移出存放到博物馆或者母国或许能够更好地实现长久保护，也能让更多人能有机会亲眼看到这些文物。也有人提出可以通过视频和照片文档的形式对文物加以保存，或许比实物形式更能达到良好的保存效果。之所以会提出这些问题，主要是因为随着科学家和游客越来越深入南极冰原以及越来越多的人工建筑在南极建设落成，南极的原始状态难免会受到影响。

历史遗址和古迹名录中的第80号遗址很有意思，该遗址为罗尔德·阿蒙森率领的探险队于1911年12月14日在南纬90°搭建的帐篷，该探险队代表人类首次造访南极点，具有独特的历史意义。现在，帐篷被埋在南极点附近的冰雪之下，自罗伯特·斯科特探险队在阿蒙森探险队抵达的一个月后到达极地时见到后，再没有人见过它。这是历史遗址和古迹名录中收入的唯一非物质遗产，虽然无法亲眼目睹，只能在图片和照片中欣赏它的样子，但并不影响它在许多人心中占有的崇高地位。

图55　1911年12月，挪威人罗尔德·阿蒙森和他率领的探险队代表人类首次造访南极点时留下的帐篷。虽然已被埋进积雪深处多年，甚至由于冰川运动早已不在最初的位置，但它依然是南极历史遗迹（第80号）。

哪些古迹和遗址已被官方历史遗址和古迹名录收录？

如想要进一步了解哪些类型的文化遗产属于南极条约体系认定应当保护的范畴，可以访问南极条约体系网站：https://www.ats.aq/documents/recatt/att596_e.pdf 。根据官方指南，申报古迹或遗址的标准有几下几条：[24]

a. 南极科考史或探索史中某一重大事件的发生地；

b. 与南极科考史或探索史中某一重要人物具有特定联系；

c. 与某件壮举或重大成就有特定联系；

d. 发展南极、认识南极的重大活动的代表或组成部分；

e. 其材料、设计或施工方式具有特殊的技术、历史、文化或建筑价值；

f. 具有通过研究揭示更多信息或让人们了解南极重大人类活动的潜力；

g. 对多国人民具有象征或纪念意义。

与联合国教科文组织世界遗产名录相同,南极历史遗址和古迹申报的基本条件是古迹或遗址应当超越地方或国家范畴,具有与南极集体历史相关的重大国际意义。用联合国教科文组织世界遗产体系的说法就是"突出的普遍价值"。由于许多古迹和遗址比较复杂(例如可能包含多类文物),其他一些文物如果都要分门别类又会导致类别数目过多,所以分析名录上的遗产类型并不容易。因此以下分析仅作概述之用,可能与其他人的分析结果略有不同。

纪念远征的古迹	18 处
纪念亡故人士的古迹	12 处
纪念早期科考站或建筑物的古迹	11 处
与英雄时代相关的古迹	28 处
国家元首访问留下的纪念物	1 处
纪念民族英雄的古迹	1 处
其他纪念馆	2 处
科考工作相关古迹	11 处
捕鲸活动相关古迹	1 处
沉船古迹	1 处
牵引车	1 处

值得注意的是,遗址总数并非92处。有2处原本单独收录的古迹和遗址后来被合二为一(第12号和第13号现被合并为第77号),3处古迹(第25号、31号和58号)由于现已不复存在,已从名录中删除。在纪念馆(不包括坟墓)类别中,或许会涉及所谓"突出的普遍价值"这一问题。过去15年间(2001—2015)增加了18处古迹。2015年,各方决定在向名单添加更多古迹或遗址前,审查整个文化遗产评判和保护体系。2018年,审查提案被提交给南极条约体系,通过审查有望改善南极文化遗产管理进程。

极地自身的文化遗产

当然，没有地方比北极点和南极点更能象征南北极了。几个世纪以来它们是探险者们追求的最伟大目标，很多人为此付出了生命的代价。

位于冰封的北冰洋中间的北极点仅是一个地理地点，覆盖其上的冰盖经常受风和洋流的影响而移动。早期，该区域过于寒冷也没有猎物，人们没有任何去那里的理由。第一批尝试去到北纬90°的人是来自北极周围国家的探险者和科学家。直到20世纪初，也不能确定在极点上或其附近是否有土地或岛屿。如前所述，1893年到1896年弗里乔夫·南森的"弗雷姆号"探险队基于测量深度的数据能够从逻辑上证明，"弗雷姆号"沿途经过的从东西伯利亚北部到斯瓦尔巴群岛的路线上（即欧亚大陆一侧）没有陆地。1926年，当挪威人罗尔德·阿蒙森、美国人林肯·埃尔斯沃思、意大利人翁贝托·诺毕尔以及机组人员驾驶飞艇"挪威号"穿越极点时，从斯瓦尔巴到阿拉斯加以及从极点到北美洲一侧，他们都没有在冰雪中发现陆地。最先声称到达北极点的有两位美国人，一位是弗雷德里克·库克（Frederick Cook）于1908年，另一位是罗伯特·皮尔里（Robert E. Peary）于1909年，然而两人的声明均无法得到证实，因而受到人们的质疑。1948年，亚历山大·库兹涅佐夫（Aleksandr Kuznetsov）率领的俄罗斯科学探险队驾驶飞机降落在北极点，成为第一支毫无争议的到达北极点的团队。

难以确定是否到达北极点的原因在前文有所提及：北极点之上并不是一片静止不动的陆地，而是一些漂浮的海冰。因此插上旗杆或者竖立纪念碑都没有作用。然而，2007年一支俄罗斯探险队以两架深潜器潜至4 300米深的海床，在此插上一面高达1米的钛合金制的俄罗斯国旗。人们认为俄罗斯的行动虽然引人注目但无法代表任何主权主张，但由于这个旗杆应该依旧竖立在北极点上，它也成为了一处文化遗产。至于公认的第一次极点探险——飞艇"挪威号"，它的出发地是斯瓦尔巴的新奥勒松。该地区保留了探险留下的一些文化遗产，包括以木材和帆布制成的巨大飞机库和铁系泊塔，这些设施建造于1925年至1926年的冬天，让飞艇在飞向极点之前可以停靠在此。

南极点似乎更容易放置永久的纪念碑，但事实并非完全如此。南纬90°的极点被约2 700米厚的冰层覆盖。由于南极的海拔高度为2 835米，除却冰层厚度，地表的海拔其实不高。覆盖海洋的冰盖并不是静止的，而是以约每年10米的速度往西经37°—40°

图56

a. 1926年5月飞艇"挪威号"从斯瓦尔巴飞往阿拉斯加时的出发地，位于新奥勒松地区。照片的前景是一个纪念碑，纪念挪威木匠费迪南德·莱因哈特·阿里尔（Ferdinand Reinhardt Arild）的巨大努力，他和他的团队建造了一个长110米、高33米、宽34米的巨大飞机库，用以在起飞前停放飞艇，飞机库建造于1925年、1926年之交的冬天。在纪念碑石块后能够看见几个支索的固定点之一，用以在风中固定飞机库。除飞机库外，在远处有一个供飞艇停泊的系泊塔。

b. 1926年5月，飞艇"挪威号"进入位于新奥勒松的飞机库。

的坐标北方向移动,向威德尔海靠近。[25] 因此,任何放置在极点上的事物都会慢慢离开极点。罗尔德·阿蒙森的帐篷就是有力的证明。1911 年 12 月这个帐篷大致位于极点之上,一个月后,罗伯特·斯科特的团队看见了这个帐篷。美国人理查德·伯德和挪威人伯恩特·巴尔肯(Bernt Balchen)在 1929 年 11 月飞越极点上空,并没有报告说见到任何帐篷。随后到达极点的是一个美国团队,他们在 1956 年将飞机降落在此,当时也没有见到帐篷。事实上在斯科特一行人于 1912 年 1 月拍摄的帐篷照片上,我们能够看到雪花在围墙四周飞舞。一段时间后,帐篷会被飘雪完全遮盖住并且压入冰层之中,同时随着冰层的移动逐渐远离极点。要想计算出帐篷如今的位置甚至将其找到需要考虑很多方面的因素,但在 2010 年进行的一次计算发现,依照冰盖移动以及雪花堆积的速度,帐篷可能在距离极点 1.8 至 2.5 千米的范围内,并且位于冰层下 17 米深处。[26] 2005 年,该帐篷被列入历史遗迹名录,以保护它不被损害,自此寻找帐篷真实位置的学术意义大于了实际意义。

所有南极洲的建筑物都是在冰层上而非在岩石上建造的,这些建筑物会逐渐被冰雪覆盖并且被挤压进冰层中,如同南极点的帐篷一样。1911 年 1 月至次年 2 月,罗尔德·阿蒙森曾在鲸湾旁的弗雷门海姆基地过冬。几年前当包裹基地的冰层碎裂成大块浮冰漂流向大海时,基地就消失在了海中。一些现代科学工作站也遭遇到同样的问题,现在建造在冰层上的新科考站采用带支架的设计,能够在需要移动的时候将整个科学站支撑起来。

在阿蒙森到达之后,罗伯特·斯科特及其四位同行者也在 1912 年 1 月到达南极点,在返回位于罗斯岛的基地时,他们五人因极寒和饥饿全部丧生。8 个月后,斯科特和另外两人的尸体在他们的帐篷中被发现,距离最终目的地还有 18 千米。尸体留在原

图 57　该图是罗尔德·阿蒙森位于南极鲸湾的越冬基地弗雷门海姆，他是第一个于 1911 年到达南极点的人。弗雷门海姆是一个大建筑群，其中的小木屋先在挪威预制好，另外还有许多帐篷，以及在冰层下凿成的隧道和挖掘成的房间。冰层下的房间可以供人们过冬，从一个房间走到另一个房间时不用遭受南极寒冷和风暴的侵袭。

地，帐篷倒塌在尸体上并且被由冰砖块堆成的巨大冰冢掩盖。和阿蒙森的帐篷一样，冰冢也会被飘雪掩盖并被埋入冰层中。如今，可能会有人推测这个墓地的位置，以及它究竟埋在多深的冰层之下。在未来的某一个时间，它可能也会因冰层碎裂流入海洋，随后被冻结入一个大冰架当中。

我们还有"极地荒原"吗

和这个问题最相关的是高纬度北极地区（高北地区），该地区的重要文化遗产能够证明原住民和外来人士在该地区有大量活动。事实上，"北极荒原"这一词语通常被理解为那些没有人类涉足的地区，这样的地区几乎不存在。在北极更广阔的地区中，几千年来都居住着原住民，来自古爱斯基摩人居住地的岩石遗迹遍布冻原，在古代狩猎或迁徙路线上的石堆能指明方向，成堆的草皮、巨大的骨头和石块表明这里曾是居住地，堆放鱼、鸟和其他动物骨头的贝丘（史前"废物堆"）表明一些小规模的家庭曾在此地停留过较长时间。通常需要训练有素的眼睛才能注意到这些历史遗迹，

图58 该图是位于加拿大威廉王岛上因纽特人的"因努伊特石堆"。在没有树木的地形上，石头以不同的方式堆放留下信息或指明方向。

并对其作出解释，这些遗迹不像其他地区的城堡和寺庙遗迹那样雄伟，但就帮助我们理解和欣赏该地区的历史而言，同样重要且不可取代。

在没有原住民的地区，例如挪威的斯瓦尔巴群岛，人类第一次开采资源是在17世纪早期，随后一批批猎人、探险家、勘探者、科学家和游客接踵而来，留下了一些废墟遗迹，如今我们认为这是值得保护的遗产，可供人们了解、欣赏，尤其是能让人了解历史。

图59　斯瓦尔巴岛的古萨马湾有着早期欧洲捕鲸活动和俄罗斯移居者波默尔人以及在此过冬的挪威猎人留下的遗迹。并且，1898年至1902年瑞典与俄罗斯联合在此处进行的大型探险活动中，俄罗斯科考站"康斯坦丁诺夫卡"有着重要地位。该探险活动旨在测量北极点附近地区的地球曲率。图6就是该次探险中的瑞典科考站。

南极大片地区有着冰雪奇观，是人类从未踏足的区域，但人口密集地区产生的污染物依旧会经空气传播影响这些地区。也有一些有人类活动的痕迹或建筑的遗迹被积雪覆盖和"吞噬"。1949年至1952年的探险期间，挪威"毛德海姆"探险队在此竖起气象仪器信号杆，1950年至1960年间，该信号杆被积雪掩盖。1956年至1960年间，挪威的"挪威站"探险队重访该地时，杆子长达

图 60
a. 一批游客排成一长列穿越西格陵兰的开阔地带。
b. 游客的鞋子踏过冻原留下的痕迹，即使在多石的路面上也清晰可见。

8米的部分都被埋在雪中，仅有几米的杆子上部露出了纯净的雪地表面。无论如何，可以想象人们总能够找到理由设法游览和重游这些地区，并对这些地区施加影响。有充分的证据表明未受影响的荒原已不存在。在这些地区，我们能看到之前所提的公认的历史遗迹，这些遗迹能够证明此处有过如勘探、捕猎海豹、鲸等历史活动。我们也能看到之后在此建造的科学站留下的遗迹，包括科学站的基地以及远离中心用于观察研究的场所。同样也有新的人类活动影响正在增加，包括通往观察鸟类和动物群落的途中，或通往文化遗迹的路线上被游客踏出的小径。

极地文化遗产不可否认的巨大吸引力

▼

即使如今游客去往极地地区是想要游览未受影响的荒原，实际上许多游览项目中远足活动的目的地依旧是历史遗迹。当然，极地游客喜欢壮丽的景色、海冰、北极熊、企鹅、冻原和冰川，但他们也会频繁地游览人类历史活动留下的遗迹，因为遗迹是极地自然环境中不可或缺的部分。人们会在书中读到这些遗迹，也会去游览。在人类更容易到达的地区中，自然和文化两者几乎密不可分。

多年以来，越来越多的游轮行将斯瓦尔巴群岛作为旅行目的地。群岛周围的遗迹位于主要定居区之外，从这些地区上岸的游客数量从 1996 年的 29 340 人增长至 2017 年的 83 571 人。[27] 如前面所提，在沿海地区分布着人类活动留下的遗迹，从 17 世纪的捕鲸活动开始，到第二次世界大战时期德国设立的一些配备有人员的气象站以及无人气象站，这些气象站建立在熊岛南部至东北地岛北部的区域。另外在内陆一些未被冰川覆盖的地区，文化遗产自然更少，但依然能够找到人类踏足的证据，那里有着勘测员留下的石堆或是勘探和科学工作留下的遗迹。

有众多网站提供去往南极的游轮旅行和其他游览，这能够告诉我们旅游业的发展使得极地地区以前所未有的方式向人们开放。对于大多数人而言，没有必要深入研究气象数据也能了解到北极正在迅速变暖并且海冰在减少。结果是，游轮现在能够去到原先只有破冰船才能安全抵达的地方。2010 年，挪威极地探险队员伯尼·奥斯兰（Børge Ousland）和三个同伴驾驶玻璃钢双体船通过东

图61　一组游轮游客聚集在东格陵兰的一处遗址附近。这是一处本土格陵兰人使用的温泉——和其他文化遗产一样，这里会因过多游客到访而遭到影响或破坏。

北和西北航道，在一个季节中完成环绕北极航行。同时，船长加夫里洛夫（Gavrilov）和全体船员驾驶俄罗斯帆船"彼得一世号"也完成了环北极航行。[28] 而在100年前，需要经特殊设计和建造的极地船只花费数年才能完成同样的航行，例如弗里乔夫·南森的"弗雷姆号"和罗尔德·阿蒙森的"莫德号"。北半球夏季时，旅游公司的游轮在高北地区行驶，随后又在南半球夏季时去到南极。并且，令人惊叹的北极经历让许多游客为之着迷，继而想要游览壮丽的南极。

　　那么现在我们要看向问题的关键部分。气候变化和不断增加的旅游业之间密不可分，这两者也对如今的极地文化遗产有着额外的影响。北极旅游业并不是新兴事物。在19世纪后半叶，一些绅士驾驶自己的船或租船驶向扬马延和斯瓦尔巴，目的是进行旅行和捕猎（除北极熊外，海象和驯鹿也是很受欢迎的猎物），更重要的是收集这些地区的地理和自然信息，因为他们记录下的一切

图 62

a. 加拿大北部剑桥湾,罗尔德·阿蒙森的"莫德号"的残骸就位于海陆交接处。

b. 图中能够看到船身坚固的木材结构,一项民间计划对"莫德号"残骸进行打捞,并于 2018 年 8 月运往挪威。

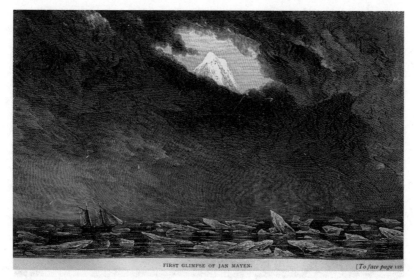

图63　该图来自达弗林侯爵的著作《高纬度来信》，该书1857年出版后成为大受欢迎的北极游记。图片展现了扬马延岛上贝伦火山山顶的景象，以及小船周围的冰使其很难登岛的场景。这也表明19世纪中叶7月中旬的气候与如今大不相同，现今岛周围没有这样的冰。

都是新信息。在达弗林侯爵（Lord Dufferin）的著作《高纬度来信》（*Letters from High Latitudes*）中，作者仅描述了一次1856年的极地旅行，就获得了巨大的成功和国际影响力。[29]

　　给没有私人船只的富人提供的"一站式旅游"迅速流行，值得注意的是一些大型的远征活动，例如由安德烈（S. A. Andrée）带领的瑞典气球探险队。1896年及1897年，安德鲁曾试图尝试从斯瓦尔巴西北部的维格翰纳海湾飞往北极点，自那时起吸引了大批游客前往该区域。1906年、1907年以及1909年期间，美国记者沃尔特·韦尔曼（Walter Wellman）也努力地尝试从维格翰纳海湾这一相同的地点飞向北极点。如今安德烈和韦尔曼的大本营的大量遗迹位于海湾边，并且自1974年起安德烈的大本营受到斯瓦尔巴文化遗产法的保护，从1992年起韦尔曼的大本营也受到该法律保护。

图64　两支想要飞往北极点的探险队留下的大量遗迹分布在斯瓦尔巴西北部的维格翰纳海湾。1896年和1897年瑞典人安德烈两次尝试用氢气球旅行，美国人沃尔特·韦尔曼在1906年至1909年期间也尝试用一艘飞艇进行旅行。探险队携带大量金属屑和废料来到现场，用以混合硫酸生产氢气，填充气球和飞艇。该遗迹是斯瓦尔巴最重要的文化遗迹之一，但因为被游客踩踏出许多小径而遭到损毁。理想情况是在海边建造一条木板路来防止进一步的损毁。

那些固定的和可移动的物品和文物不应被移动、损害或取走。弗里乔夫·南森是挪威北极科学家以及弗拉姆探险队的队长，他在1893—1896年间穿越了北冰洋。在1920年去往斯瓦尔巴的科学活动中，他乘坐自己的游艇游览维格翰纳海湾，且需注意的是（此处为作者的译文）：[30]

> 大多数有用及有价值的物品，尤其是金属制品，我认为如今应该已被盗走，但也有一些东西留存了下来——捕猎者和游客无法全部拿走的东西。（p.145）

> 另外，游客们也会到来这里，把他们的名字划刻到各个角落，以此作为纪念。（p.146）

尽管，在过去和如今，驾驶私人游艇驶往南极都是寻常的事，但这在极地旅游业中只是较小的部分。20世纪20年代，一些船为南极诸岛上捕鲸和捕海豹的工作站提供服务，可以在这些船只

// 极地文化遗产

图65　照片为2006年摄于北极的极地游轮"林德布拉德探索者号"。尽管从1969年开始她始终安全地在北极和南极航行,但2007年,在南设得兰群岛附近,她因撞上坚冰而沉没。船上的91名乘客以及63名船员和向导被疏散到救生船和充气小艇中,非常幸运的是当时海面平静且天气很好,在第一艘救援船到达时,所有人已在海面上漂浮了约5小时。船在撞上坚冰约20小时后沉没。

上购买游客床位。第一次去往南极的商业游轮旅行由瑞典裔美国企业家拉尔斯-埃里克·林德布拉德(Lars-Eric Lindblad)在1969年组织进行。作为去往南极和北极的先驱船,他特别设计的船只"林德布拉德探险者号"成了著名的大型游轮。不幸的是,该游轮在2007年11月重获声名的原因是它撞上坚冰,成为第一艘在南极洲沉没的大型游轮,幸运的是没有人员丧生。

2017年到2018年的南极旅游季中,43 691名游客踏足南极,而前一年旺季仅有不到37 000名游客。[31] 43 691名游客中很大部分来自两个国家:13 412名来自美国,7 881名来自中国。从南极大陆的面积来看,这些游客看起来可能并不多。然而,必须注意的是南极的夏季很短,并且能够让游客登陆游览的遗迹数量也相对较少,因此游客会频繁地造访这些遗迹。另外,旅游季也和南极动物的繁殖季重合。在北极一些类似的自然或文化遗产处,有很多游客在附近走踏,会不可避免地产生负面影响。

对北极文化遗产的威胁

气候变化正在威胁着北极文化遗产，海蚀和更加温和、湿润且更多风暴的天气破坏着那些原本被干燥和寒冷的气候所保护的遗迹。人们会因有关北极熊数量减少和生存受到威胁的报导而感到沮丧，实际上，保护和管理文化遗产的工作似乎也同样令人沮丧。

长久以来人们普遍认为北极文化遗产会被"冻结在过去"，但随着气候变化的影响，如今遗迹正受到严重的影响。1987年，当一本有关尸体解剖的书出版时，这一"冻结在过去"的原则变得非

图66　图中是位于加拿大北部比奇岛的坟墓，坟墓的主人是约翰·富兰克林爵士西北航道探险队中的三名成员。1845年至1846年，这三名男性死于探险旅程中的第一个冬天。两艘船只和其余126名男性全部消失在加拿大本土附近无人涉足的岛屿中，一度行踪难觅。此后很多年中，这三座坟墓成了这次探险的唯一证明。在永冻层中，尸体的保存近乎完好。

常有名。书中记述的是1984年和1986年分别对两具埋葬在比奇岛的尸体进行的解剖,死者逝于1845年约翰·富兰克林爵士西北航道探险的灾难中。

在被埋葬140年后,我们依旧能够辨认出尸体和他们的衣物,并且能够将头发和软组织进行取样分析。[32] 海冰的缺乏导致更多的海岸侵蚀,融化的永冻层影响建筑地基并且让原本被埋藏的有机物质暴露出来而遭降解,更多的霉变和腐烂使木头毁坏,更多的暴风雨天气损坏了脆弱的建筑,从这些方面均能够看出是更为温暖、潮湿和更多风暴的北极气候对文化遗产造成了负面影响。并且,上述所提的旅游业增长也是造成负面影响的因素。

斯瓦尔巴最重要的100处受法律保护的文化遗产无一例外都位于沿海地区。[33] 同样,北极的许多遗迹也都是如此。这是物流条件和地理环境导致的结果:如果能够在沿海地区寻找到合适的

图67 该图展示的是一处1957年建于斯瓦尔巴群岛的房屋中发生的腐烂和霉变。该房屋是受法律保护的历史科考站10处建筑之一,已经多年无人使用。尽管斯瓦尔巴东北部的天气寒冷干燥,但建筑物内部的微气候给腐烂和霉变提供了环境。如今出于健康考虑,不建议人们进入该建筑或在建筑内停留。理想状况下,应该通过使用该建筑的方式对其进行保护,但首先需要进行大范围的修复工作。

资源，人们就不会继续去往内陆。原本在夏季的大多数时候和整个冬季都附着陆地边缘的冰能够作为屏障防止海浪侵蚀。然而，海冰持续融化，即使在冬天也是如此，这样的情况使得屏障消失。另外，由于北极的风暴天气增多，海浪侵袭也更加频繁，因此北极所有沿海地区都受到更多侵蚀的威胁。[34]

图68 在作者和女儿背后的左方能够看到一圈陆地边缘的冰外缘线。该照片摄于1982年，在过去的岁月中，斯瓦尔巴的许多峡湾缺乏海冰的屏障，陆地边缘的冰外缘线则变得更加罕见。在冬天，海面上的冰受到风浪拍打而碎裂，当海岸上的冰比海面上的冰留存更久时，海岸边会有残留的海冰层。冰外缘线的存在能够帮助保护海岸不受海浪侵蚀，现在海冰的减少导致北极海岸遭到侵蚀的情况越来越严重。

因此，海岸线后撤至越来越接近文化遗产的位置，最终文化遗产也将被侵蚀并掉入海中。当脆弱的冰融化，这些在地面沉积物中起黏合作用的冰消失而导致地面"塌陷"时，在有更大冰层或冰透镜体的永冻层中，侵蚀作用就会大大加速。在大量高含冰量的永冻层中，例如加拿大西北部地区的塌陷面积达到了40 000平方米，有高达25米的冰斗。[35] 塌陷会严重影响现代和传统的居住

图 69　照片为两幢砖楼，位于斯瓦尔巴金字塔镇废弃的俄罗斯矿业定居点。砖混结构建筑不如木质建筑柔韧，永冻层的运动会使得建造在其上的砖混结构建筑产生开裂和不稳，可以在入口楼梯左边的窗户周围看见这些裂痕。融化的永冻层正在给北极所有的基础设施带来严重的问题，例如使道路、简易机场跑道、管道以及建筑变得不稳。

图 70　斯瓦尔巴群岛朗伊尔城一个较大的挪威城镇中最古老的建筑之一。与潮湿的地面接触使得木制墙壁靠近地面的部分开始腐烂。当 2007 年拍摄该照片时，该建筑正在进行一次全面的大整修。在很多案例中，必须将这样的建筑抬高远离地面。一些特殊的建筑可能无法抬高，但面对天气更加温和潮湿的现实情况，这也是我们能够作出的妥协。

图71　1852年至1853年英国海军部组织了一次搜索行动，追踪去寻找西北航道的约翰·富兰克林爵士所带领的探险队的行迹，海军在加拿大比奇岛建造了"诺森伯兰小屋"，该建筑使用的建材是打捞上来的捕鲸船的桅杆和其他木材，并且搜索人员抱着徒然的希望，认为富兰克林探险队的一些成员可能依旧活着。1845年至1846年，探险队成员曾在该岛屿过冬，搜索人员认为他们可能会返回岛屿。这座小屋储藏了能够帮助幸存者的物资。然而，富兰克林探险队的成员早已死亡，并且这座小屋中的物品和用具后来被他人掠夺，以及因气候而损毁。如今，小屋内外的木材和用具以及小屋周围的其他遗迹看起来只是垃圾，但每一件物品都有一个故事，能告诉人们它们的过往。游客很难理解为什么不让他们从这些遗迹上走过，因为每一个脚印都会进一步伤害这些脆弱的遗迹。对于极地地区那些规模不大的历史遗迹而言有一个共同的问题，那就是未受训练的双眼无法从这些看似垃圾的遗迹中发现它们的历史价值。

区及运输路线。另外，融化的永冻层会使得楼房和建筑物地基不稳，给文化遗产增加新的压力。

捕猎者、勘探者、探险者、以及其他人在北极留下了许多简单却历史悠久的重要木制建筑，这些建筑都直接建造在冻原上。当气候变得更加温和湿润时，木头由于腐坏和霉变而损坏。这未必是新问题，但在新的气候条件下腐坏和霉变会加速发展。

并且，那些原本作为屏障的海冰现在已消融后退，那些几十年以及几个世纪没有人到访的遗迹现在逐渐成为个人和团体的游览目的地。大多数游客当然不想给遗迹造成负面影响，但他们周围的遗迹、植被和地形都非常敏感，即使只有个别游客走过也会无意踩到一些对遗迹有保护和稳定作用的小植物，并且，游客走过会压坏那些已经老化的木制建筑遗迹和文物。另外，少数游客对于他们造成的伤害显然毫无察觉或漠不关心，可能会在遗迹处涂鸦，非常不小心地对待文物，甚至再带走一些"纪念品"。

图72

a. 比起对历史遗迹的无意踩踏，更糟糕的是游客故意破坏，这通常是由于缺乏对于遗迹的理解。几年前，这样的涂鸦损毁了一个17至18世纪非常重要的墓地，该墓地位于斯瓦尔巴的西北部。

b. 最近一次在斯瓦尔巴历史遗迹上出现涂鸦是在2016年，一名游客在朗伊尔城的最大城镇中几处遗迹和其他建筑上涂鸦留字。

文化遗产的负面影响有助于解释气候变化

▼

关于气候变化的事实已在前文多有提及，并且通过不同自然科学学科收集的详细信息通过政府间气候变化委员会和许多其他渠道传播给大众。然而，除了自然科学界进行的大量观察和测量工作，人类也可以通过历史学、考古学、历史考古学等 学科以及与物质遗产相关的工作来揭示和证实气候变化问题。

历史能够告诉我们一幢房屋或建筑建造于何时，并且有可能根据当时的景色提供更多有关建筑情况的细节，这可能有助于记录海蚀的情况。例如，一位科学探险队成员的日记中可能提及帐篷搭建在距离海岸多远处，一张拍摄勘探或矿业居住点的照片可能也会展示出同样的信息，之后使用这些房屋或建筑的人所写的日记可能也会给出有关侵蚀速度的线索。例如，斯瓦尔巴的捕猎站"弗雷德海姆"建造于1927年，位于远离海岸的安全位置，捕猎站由一幢主屋和两幢小建筑组成。对侵蚀速度的测量从1987年开始，当时主屋距离海岸边缘的距离为17.7米。到2011年该距离已经缩短至8.74米。[36] 早在2001年，建筑群中最老的一幢小屋距离侵蚀边缘仅有3米的距离，显然已有坠入大海的危险，于是小屋被向后移动了6米。2012年，测量显示主屋距离边缘的位置有8.5米，到2014年时该距离仅有6米。唯一能够拯救遗迹的办法是将其移动，2015年4月，整个捕猎站被向后移动了37米，远离海岸线。

扬马延是挪威的一座北极岛屿，位于大西洋洋中脊的最北端，

图 73　位于斯瓦尔巴的捕猎站"弗雷德海姆"由一名为希尔马·诺维斯（Hilmar Nøis）的挪威捕猎者于 20 世纪 20 年代建造，他在斯瓦尔巴度过了 35 个冬季，其中 19 个冬季在捕猎站度过。他的捕猎目标是北极狐皮。他的故事十分出名，弗雷德海姆也因此在斯瓦尔巴的遗迹中有着特殊地位。这里的海岸线受海蚀的影响越来越严重，2015 年不得不将整座捕猎站移动到山脊上，照片左侧的旗杆附近就是工作站。

是北美大陆板块和亚欧板块与非洲板块的交界处。熔岩在裂谷带不断涌出，洋中脊周围岛屿的成因都与此有关。扬马延海岸附近的海域具有很强的侵蚀力，尤其是在有着松动的火山沙及火山碎屑的地区。1615 年至 1645 年期间，海滩上有大规模的捕鲸活动，荷兰捕鲸者肢解鲸的尸体并从鲸脂中煮油。历史图片中展示着广阔的沙滩、煮鲸脂的炉子和一些捕鲸者在海岸边使用的建筑。历经几个世纪后这些海滩被冲蚀殆尽，这些 17 世纪的建筑几乎没有留下任何痕迹。1930 年，荷兰海军舰艇的船员在距离海岸线 80 米处的沙滩内坡放置了一块重达 500 千克的纪念石，这些在照片和书面报告中都有记录。到 2014 年夏天，这块石头距离大海仅有 21 米，并且内坡和石头都处于一起崩塌跌入大海的危险之中，因此该纪念石被移去了海湾西部更为平坦和安全的区域。历史资料证明该海滩宽度由于海蚀在1930年至2014年间减少了60米。1980年,作者第一次去到该海滩时,能够明显看见捕鲸时代最中心的建筑遗迹,但如今所有遗迹都已消失在大海中。

图74　照片为扬马延岛的弗雷罗斯布科达海湾,摄于2014年。该海湾是17世纪上半叶荷兰捕鲸活动的一个重要遗迹,在20世纪90年代能看到工作站留下的许多遗迹。当时人们在此处从鲸脂中熬油,但如今几乎所有的遗迹都被侵蚀消失在海中。箭头所指的纪念石在1930年被放置于距离海岸线80米处,到2014年时距离大海仅有21米,内坡也有全部崩塌的危险,纪念石因此被移动。

在其他极地探险报告和日记中,有着有关气象条件、海冰、植物群和动物群的丰富信息。在早先还未进行长期测量和观察的时候,这些信息有助于了解该时期的气候条件。海员和捕鲸者的航行日志及日记帮助拼凑出了北极海冰的历史状况,在一篇题为"拼凑北极海冰历史,记录可追溯至1850年"[37]的文章中,美国国家冰雪数据中心首席调查员弗洛伦斯·弗特来(Florence Fetterer)叙述了如何通过一系列资料帮助填补北极海冰记录的空白以及建立更多新记录,并将至1850年的海冰记录补充完整。这些资料包括捕鲸船的航海日志和北大西洋在1850年至1978年间的海冰外缘线位置，位置资料则来源于报纸、船舶观测、飞行器观测以及日记等各种出处。

考古学从永冻层的变化入手进行观察。与上述提到的1846年比奇岛坟墓调查一样，1980年考古学家对位于斯瓦尔巴西北部的17世纪捕鲸人坟墓进行了考古发掘，发现尸体仍保有皮肤和头发，他们的羊毛衣物保存完好，几乎还可以穿。在2016年及2017年，在同样的区域发掘出相似的坟墓，但由于永冻层的位置下降，无法再将物体"冻结在过去"，坟墓中几乎没有留下任何东西。

图75

a. 图中的衣物来自17至18世纪的捕鲸者，他们位于斯瓦尔巴西北部的坟墓在1980年被发掘。即使是浅埋在永冻层中的坟墓，其中的衣物也能保留下来，在一些案例中尸体的皮肤和头发也得以保留。

b. 在2016年对同一地区进行挖掘时，永冻层已下降至棺材以下的位置，坟墓中几乎没有留下任何衣物、皮肤或头发。

同样，永冻层融化也正在损害西格陵兰贝丘中的有机物质，这些有机物质中包含的证据能证明3 500年前的三个主要格陵兰文化的存在——萨卡克、多塞特和图勒。[38] 意识到这种独特的考古学材料会在80到100年间完全消失，丹麦研究永冻层的科学家提出进行专门研究。他们的研究显示，那些通常会吞噬有机物质（例如木材、骨头、软组织等）的细菌处于永冻层中时会休眠。一

图76 该图是位于努克的格陵兰国家博物馆与档案馆,馆中藏有来自格陵兰各处的藏品,目的是保护和推广格陵兰的历史。藏品和展览所涵盖的历史范围很广,包括第一批到达格陵兰的文明,4 500年前的萨卡克文明以及1475年左右埋葬的六名女性和两名男孩的干尸。由于寒冷干燥的条件,1972年尸体被发现时,其穿着的衣物和被包裹在内的皮肤保存得十分完好。

旦永冻层融化,这些细菌就会变得活跃,并且在这个过程中会产生热量,使更多的永冻土融化——能够发起这个有趣的研究正是有赖于考古学家和永冻层科学家间的相互交流。

用科技保护文化遗产

修理和修复是保护楼房和建筑并延长其寿命的传统方法。在极少数案例中,也会使用移动遗迹的方法,例如移动一幢受到侵蚀威胁的小楼房,如之前提及的位于斯瓦尔巴的"弗雷德海姆"捕鲸站采用的就是该方法。监控自然影响造成的后果(例如木材的侵蚀与损害)和监控人类活动造成的影响(例如对历史遗迹和周围植物的损伤)是保护文化遗产的重要方法,并且会持续开展下去。

除却传统方法,人们愈发注意到可以在相关领域利用新科技去保护遗迹。无人机能够被用于监控遗迹和记录变化,例如监控侵蚀程度增加时,操作员不必亲自踏足遗迹处。例如,使用无人机监测加拿大西北地区的沿海地区迅速增加的侵蚀[39],以及对斯瓦尔巴西北部遗址进行现场调查。

另外,监测卫星的发展覆盖到北极区域,这开拓了令人激动的领域,使得远距

图77 这是一张由无人机拍摄的合成照片,展示了斯瓦尔巴群岛西北部 17 至 18 世纪捕鲸人的墓地。大海和海滩位于图片的底部。中间的许多小椭圆是一些分布在四处的坟墓,用石堆覆盖以保护尸体免受北极熊和狐狸的伤害。坟墓周围和坟墓之间的浅色区域是由于游客踩踏而磨损的本就稀少的苔原植被。这也表明了为什么需要在文化遗迹分布的主要区域限制旅游业的发展。

离的信息收集成为可能。欧盟哥白尼计划对该领域很有兴趣。它旨在基于卫星对地观测和实地（非空间）数据来发展欧洲信息服务。[40] 引入遥感工具开辟了监测偏远地区文化遗产的新领域，并且提供了全新的机会，能在不造成附带损伤的情况下对这些遗迹进行更多研究，而传统的探险方法会对这些遗迹造成不可避免的损害，包括空运和海运的排放以及人类对遗迹的直接影响。

另一先进的科技也被引入到遗迹保护中，那就是遗迹专家乐于使用的扫描技术。详细的测量、照片、比例图和书面描述是对历史遗迹进行文件记录的主要传统方法。因为遗迹可能会在日后被损毁，例如经历侵蚀、火烧或者其他的自然损害而不复存在，为让遗迹在记录中得以保留，就需要非常仔细和精确地进行记录，也意味着在收集资料时需要花费很多时间并投入其他资源。在很短的时间内，仅需两个操作员就能使用3D激光扫描将非常复杂的

图78

a. 该图是20世纪后，亚南极南乔治亚地区的捕鲸活动所留下的巨大工业建筑群。这是非常重要的文化遗产，但很难将遗迹中的一切尽数保护起来。

b. 因此，英国南乔治亚行政当局在挪威的支持下，对其中4个主要工作站的内部和外部进行了全面详尽的3D扫描记录。这份详尽记录可以视作一张空中拍摄的照片，但扫描能将地面上获得的一手材料也记录其中。通过这份文件记载的信息，能够实现对工作站的虚拟游览，通过研究独立扫描或动态扫描就能够让参观者"游览"一些遗迹和楼房。

/93

文化遗迹捕捉下来。诚然，之后分类并展示收集到的数据仍旧需要时间，也需要专家、合适的软件和电脑技术，但这些工作都能够在办公室中进行，需要在实地进行的工作大大减少，效率也得到提高。作者曾在亚南极的南乔治亚地区参与到一次扫描工作中，对一个复杂的旧工业捕鲸站进行整体扫描，仅需两名操作员在现场花费几天的时间就能对整个工作站进行从内到外的全面记录。

如之前所提，可以在网站：https://www.youtube.com/results?search_query=geometria+ltd 上找到更多例子，并且以此种方法保存下来的文件记录不仅能够让游客对历史上的捕鲸站进行虚拟游览，也能够为不同的研究项目所使用，例如和工作站分布及建筑风格、土地使用和捕鲸历史有关的研究。另外，研究房屋和建筑的细节或许能有助于保护和修复一些建筑的特殊部分。

该技术也同样在高北地区使用。2010年，一台激光扫描仪被用于扫描康格堡，该历史遗迹位于加拿大埃尔斯米尔岛的富兰克林夫人海湾。关于该项目的文件中记述道：

图 79 南乔治亚的废弃捕鲸站充斥着石棉。出于健康因素，工作站周围 200 米禁止人们接近。2015 年，作者与英国政府一起参与了对工作站的一项调查。根据规定，我们必须身着全身防护衣来防止石棉尘的危害。一直佩戴着头盔和面罩让我们感觉很不适！

由于气候变化、气象、野生生物和人类活动的影响，康格堡正处于危险之中。在该文件中，我们展示了在严酷的环境下，3D激光扫描是如何快速准确地记录遗迹的文化特征。我们商讨着通过使用3D扫描数据作为基线，能够怎样管理自然进程和人类活动在未来造成的影响，怎样通过得出的模型来计划保护和修复工作，以及怎样利用通过激光扫描数据创建的3D模型来激发公众对于文化管理和北极历史的兴趣。[41]

该文件对于激光扫描科技的使用进行了相当出色的描述，该科技能够应用于所有大小、所有类型的物品和遗迹。

怎样减少旅游业的影响

▼

旅游业对高北地区和南极的影响不断增加,这使得遗迹管理者不仅需要采取行动引入管理规章和限制,也要尽可能向游客和旅游业从业者展示关于不同历史遗迹的大量信息。一旦游客被告知或读到一些信息,了解到某处遗迹处发生过的重要历史事件,或者了解到遗迹经历了比如今更为严酷的气候条件后得依靠人们坚定地采取非常措施才得以保存,就会对保护表示理解和赞同。绝大多数情况下,游客在听到这样的例子后会对遗迹保持敬意和关切,只会拍摄一些照片作为纪念,将当下的独特体验和久远的历史故事一同保留在记忆当中。

在人们明白需要了解相关遗迹信息,也需要将其告知他人的情况下,极地地区各种历史遗迹的信息不断增加,为描写这些地区历史的畅销书籍提供了素材,也希望这些书籍能够大大增加人们对于极地地区的兴趣。然而,有很多方面需要仰仗开往极地地区的游轮上的旅游从业者、船员和向导。他们既有责任了解相关的法律和规定,也有责任从道德方面给予指导,让游客以对环境最为友好的方式参观那些脆弱的地区,他们可以告知游客并进行仔细指导。很重要的是,人们需要明白有些地区和遗迹完全不应该成为旅游景点。因为任何人类影响都会造成负面后果,例如,侵蚀加剧,或者进一步损害有稳定遗迹作用的植被和散布各处的文物。

许多运营南极大型游轮的公司都是国际南极旅游组织协会的成员,该协会网站主页为:https://iaato.org/home。在北极类似的组织是北极探险旅行组织协会,该组织网站主页为:https://www.aeco.no/。除却作为旅游公司的成员组织,这两个机构也提倡进行负责任的极地旅行,通过制定指导方针来告知从业者最佳实践方法,并且在游览文化自然遗址、野生生物和当地居住点或科考站时,尽可能地减少对它们的负面影响。这是商业中的必要做法,也受到了地区管理者的大力支持。不幸的是,并非所有从业者都是两个组织的成员,私人探险队也显然不在其中。

旅游从业者和其他人员时常会提出一个观点,那就是文化遗产属于全人类,因此应该向所有人开放以供游览。以下的事实能够反驳该观点,全世界具有国际重要性的一些文化遗产目前都处于限制参观或完全关闭的状态。例如,在联合国教科文组织认定的世界遗产中,智利的复活节岛石像、埃及国王谷的法老王陵墓以及秘鲁的马丘比丘印加城都因过度游览而遭到损害,那些曾使它们闻名的细节也因此遭到破坏,这些遗产正在实施或已实行了游览限制规定。法国拉斯科洞穴内有一些旧石器时代洞穴壁画,已有超过 20 000 年的历史,是迄今发现的最令人惊叹的洞穴壁画。当人们发现参观洞穴时人类的体温、湿度和二氧化碳排放会对遗迹造成很大伤害时,就在 1963 年关闭洞穴不再开放给公众参观。20 年后,一个"仿真洞穴"在遗迹附近开放,以满足公众的参观需求。如今,通过虚拟现实体验可以做到很多事。埃及国王谷中少年法老图坦卡蒙的坟墓在 2010 年至 2014 年间关闭进行修葺,关闭的原因是自从 1922 年坟墓被发现之后,数以百万计的参观者到该遗迹处游览,对其造成了损害。2014 年 5 月,一间按照具体细节进行全面重建的墓室开放给公众参观。尽管原遗迹已对公众开放,但出于保护遗迹的原因必须考虑将其关闭。因此,关闭重要的极地文化遗迹或限制游客数量的做法也并非特例,相反,这是

图80　哈尔·萨夫列尼地下宫殿是联合国教科文组织认定的世界遗产，位于马耳他，是一处包含楼梯和房间的大型地下建筑，已有超过 5 000 年的历史。因游客参观造成的湿度上升使得遗迹遭到损害，这让政府决定严格控制进入该遗迹的参观人数。

如今国际文化遗产管理中的基本办法。事实上，在斯瓦尔巴已经采取了类似做法，有 9 处非常重要的历史遗迹已经在 2009 年采取了限制参观的保护措施。

对于未来的考量

▼

从 1979 年开始，过去的 40 年中，作者专门在极地地区从事相关工作，这让作者能够很容易地证实一些变化已经发生。从前，海冰和陆地上的雪使得在斯瓦尔巴的实地工作只能在每年夏季的 2 个半月至 3 个月中进行，全年中只有 7 月和 8 月会有游轮去到极地地区，其他时候很少有游轮到来。如今，在 3 月底就有短期航行的游轮出发，旅游季到 10 月才结束。换言之，如今限制旅游的因素已经变成了白天的长短而非海冰的阻碍。

极地地区的景色十分壮丽，人们希望能够到极地游览的心愿合乎情理。新闻经常报道有关极地冰雪区域减少的消息，以及警告人们北极和南极陆地边缘的冰正在融化，这会推动人们前往极地旅游。有一种想法是人们必须在极地地区消失之前去游览，并且通过这种方式，帮助我们加强对气候变化的影响力。据说去往极地地区的旅行非常令人动容，人们在游客当中选出了一些大使来宣扬极地保护。然而，带更多人去极地体验，甚至让一些游客成为大使来宣传极地保护，或者是杜绝人类活动以免对极地产生影响，这两种保护遗迹的方式之间存在差别。近年来的营销概念是推广"生态友好"，参加对生态有利的旅游航行去"见证"北极的气候变化以及因变化而遭受痛苦的北极熊和海豹，但这种说法忽略的事实是，一艘游轮每天排放的二氧化碳相当于 13 000 辆柴油汽车的排放量，更不必说其排放的大量大气颗粒物和氮氧化物污染物。[42] 更不为人注意的是，游轮的引擎噪声会对海洋生物造成影响，也会影响那些被海冰覆盖的地区，船只破冰前行打开航道的行为会侵扰野生生物，也会破坏海冰的白色表面使得深色表面吸收更多的太阳热量，进一步加剧温室效应。毫无疑问，游轮的拥有者已经能够看见"不好的预兆"，未来，他们必须重新设计轮船来应对并缓解这些负面影响。采取这些行动需要时间，但希望对新游轮实行的有关燃料种类和污染控制的规则和改进会在一定程度上使情况逐渐得到改善。

图81　盖朗厄尔峡湾是联合国教科文组织认定的世界遗产，陡峭的山从两侧将峡湾包围，秀丽的风景使得人们十分喜爱乘坐游轮到此游览，总能看到峡湾中同时有好几艘大游轮行驶。游轮排放产生污染，破坏着风光美景和居民的健康。

我认为必须接受的一点是，高北地区以及南极的大部分区域应该禁止团队旅游，有些区域则应该完全禁止游客进入，其他的一些区域或许可以允许个人或小团队在严格遵守指导方针的情况下进入游览，将造成的影响减至最小。实行这样严格的措施去保护鸟类或动物，让它们能够继续生存繁殖，这似乎更容易为人们接受。而文化遗迹并不能繁殖再生，人们倾向于认为文化遗迹完全开放给所有人是人类的某种权利。早在1973年，斯瓦尔巴东部的卡尔王地群岛全年禁止游客进入，仅允许得到批准的相关人员偶尔进入进行一些科学考察，这是为了保护北极熊挖掘洞穴进行生育的地区。人们从未强烈质疑禁止游览的规定，因为我们都能够理解这是出于保护北极熊的需要。同样，位于斯匹次卑尔根岛北部，大小约2千米×3千米的莫芬岛被设立为自然保护区，以保护离开水域前来此处觅食和休息的海象。非常奇怪的是，关闭一些区域来保护重要的文化遗产就会遭到更多更强烈的反对。我们可能需要回忆几十年以前的思维方式，当时那些令人惊喜的电视和电影纪录片是深入了解那些我们无法踏足的区域的唯一方式，例如深入丛林、雨林、广阔的沙漠、冰封的南极，那时的我们能够接受这一点。直到有技术能够让我们真正去游览那些地区，但同时又不会给周遭的一切带来负面影响之前，虚拟现实技术或许能够帮助我们拯救那些至关重要的极地文化和自然遗产。

图82 格陵兰西部的加达是12至14世纪格陵兰岛古斯堪的纳维亚人定居点的主教座堂所在地。该遗迹靠近如今的伊加利科居住区。许多来自古斯堪的纳维亚教堂的巨大石块和主教的住所依然保留在原处，而其他建筑已经融入新的楼宇之间。该遗迹位于格陵兰岛相对富饶的区域，有着令人赞叹的美景。

/ 101

注 释

1. https://www.britannica.com/event/Paleolithic-Period
2. https://www.britannica.com/science/human-evolution
3. http://www.historyworld.net/wrldhis/PlainTextHistories.asp?historyid=ab25
4. http://www.historyworld.net/wrldhis/PlainTextHistories.asp?gtrack=pthc&ParagraphID=bei#bei
5. https://www.britannica.com/topic/Peking-man
6. de Caprona, Yann 2013: Norsk etymologisk ordbok. Kagge forlag AS, Oslo. 此处引文由作者翻译转述。
7. http://www.unesco.org/new/en/culture/themes/illicit-trafficking-of-cultural-property/unesco-database-of-national-cultural-heritage-laws/frequently-asked-questions/definition-of-the-cultural-heritage/
8. 特此感谢美国华盛顿史密森尼北极研究中心的威廉·菲茨休（William Fitzhugh）主任对本段作出的贡献。
9. http://pbsg.npolar.no/en/agreements/agreement1973.html
10. Susan Barr, David Newman and Greg Nesteroff 2012: Ernest Mansfield, Gold or I'm a Dutchman. Biography. Akademika Publishing, Trondheim, Norway.
11. http://www.gov.gs/
12. Barr et al. 2013: Assessment of Cultural Heritage Monuments and Sites in the Arctic. Arctic Council (SDWG) Project #P114:2
13. 同上 :6, 7
14. 同上 :7
15. https://lovdata.no/dokument/NL/lov/2001-06-15-79?q=svalbard
16. https://www.riksantikvaren.no/Veiledning/Data-og-tjenester/Askeladdenand Kulturminneplan for Svalbard 2013-2023, Sysselmannen på Svalbard Rapportserie Nr.1/2013, p.31.
17. http://dqbglhnbfffy1.cloudfront.net/fileadmin/user_upload/Inatsisartutlov_nr_11_af_19_maj_2010_om_fredning_og_anden_kulturarvsbeskyttelse_af_kulturminder.pdf
18. http://pubs.aina.ucalgary.ca/arctic/Arctic57-3-260.pdf
http://ipy.nwtresearch.com/Documents/Hunters%20of%20the%20Alpine%20Ice.pdf
19. https://www.nps.gov/gaar/learn/historyculture/landscape-archaeology-at-agiak-lake.

htm

20. https://web.archive.org/web/20090113130338/http://www.natmus.dk/sw18660.asp

21. https://www.nzaht.org/explorer-bases/scotts-hut-cape-evans

22. https://www.nzaht.org/pages/history-of-the-projects

23. 更多信息可见 https://www.tandfonline.com/eprint/KnEmZnK4YazRB9b4vXS7/full

24. https://www.ats.aq/documents/cep/Guidelines_HSM_V2_2009_e.pdf

25. 参见网站：https://en.wikipedia.org/wiki/South_Pole#Ceremonial_South_Pole

26. 《极地记录》，第47卷，第3期，2011年6月。

27. 参见网站：http://www.mosj.no/no/pavirkning/ferdsel/cruiseturisme.html

28. 参见网站：https://www.seilmagasinet.no/innhold/?article_id=31015 和 http://www.yachtingworld.com/news/world-record-for-russian-crew-in-arctic-7471

29. Dufferin, F.H.T.B. 1857: Letters from High Latitudes. John Murray, London.

30. Nansen, Fridtjof 1920: En Ferd til Spitsbergen. Kristiania.

31. 数据来自国际南极旅游组织协会。

32. Beattie, Owen & Geiger, John (1987). Frozen in Time: Unlocking the Secrets of the Franklin Expedition. Saskatoon: Western Producer Prairie Books. ISBN 0-88833-303-X.

33. https://www.sysselmannen.no/globalassets/sysselmannen-dokument/trykksaker/katalog_prioriterte_kulturminner_paa_svalbard_versjon_1_1_2013_komprimert.pdf , page 8

34. 参见网站：https://www.livescience.com/13746-arctic-coast-erosion-climate-change-ice.html

35. 参见网站：http://www.nwtgeoscience.ca/project/summary/permafrost-thaw-slumps

36. Note 61, p. 71 in Kulturminneplan for Svalbard 2013-2023, Sysselmannen på Svalbard Rapportserie Nr.1/2013.

37. 参见网站：https://www.carbonbrief.org/guest-post-piecing-together-arctic-sea-ice-history-1850

38. 参见网站：http://sciencenordic.com/climate-change-destroying-greenland%E2%80%99s-earliest-history

39. 参见网站：http://www.cbc.ca/news/canada/north/drones-monitor-nwt-arctic-shoreline-erosion-1.3897042

40. 参见网站：http://www.copernicus.eu/main/overview

41. 参见网站：http://polarheritage.com/content/library/Arctic_Application_of_3D_Laser_Scanning.pdf

42. Reported in the newspaper Klassekampen 19 July 2018.

图片来源

图1：苏珊·巴尔
图2：苏珊·巴尔
图3：理查德·巴尔（Richard Barr）
图4：苏·多诺万（Joe Donovan），https://commons.wikimedia.org/wiki/File:Harvesting_the_wheat_-_geograph.org.uk_-_144209.jpg
图5：
a. 雷特·巴特勒（Rhett Butler）
b. 韦提希·沙（Vaidehi Shah），https://commons.wikimedia.org/wiki/File:Litter_on_Singapore%27s_East_Coast_Park.jpg
图6：苏珊·巴尔
图7：
a. 维基百科
b. Spyrosdrakopoulos, Wikimedia, https://commons.wikimedia.org/wiki/Category:Zollverein_Coal_Mine_industrial_complex#/media/File:1417_zeche_zollverein.JPG
图8：
a. 马库斯·托马森（Marcus Thomason），弗雷姆博物馆
b. https://commons.wikimedia.org/wiki/File:U_240,_Lingsberg.JPG
图9：苏珊·巴尔
图10：苏珊·巴尔
图11：苏珊·巴尔
图12：苏珊·巴尔
图13：米斯蒂·切尔诺夫（Mstyslav Chernov），Wikimedia, https://commons.wikimedia.org/w/index.php?search=Tourists+Mykonos+Island+Greece&title=Special%3ASearch&go=Go&ns0=1&ns6=1&ns12=1&ns14=1&ns100=1&ns106=1#/media/File:Little_Venice_quay_flooded_with_tourists._Mykonos_island._Cyclades,_Agean_Sea,_Greece.jpg
图14：中华地图学社绘制
图15：中华地图学社绘制
图16：苏珊·巴尔
图17：中华地图学社绘制
图18：公共领域
图19：中华地图学社绘制
图20：苏珊·巴尔
图21：公共领域
图22：苏珊·巴尔
图23：公共领域
图24：苏珊·巴尔
图25：苏珊·巴尔
图26：
a. 苏珊·巴尔
b. 厄于斯泰因·威格（Øystein Wiig）
图27：苏珊·巴尔
图28：中华地图学社据挪威极地研究所所绘图片重新绘制
图29：*Ernest Mansfield, Gold or I'm a Dutchman* by Susan Barr, David Newman, Greg Nesteroff, 2012
图30：苏珊·巴尔
图31：苏珊·巴尔
图32：苏珊·巴尔
图33：苏珊·巴尔
图34：苏珊·巴尔
图35：苏珊·巴尔
图36：苏珊·巴尔
图37：马库斯·托马森（Marcus

Thomason），弗雷姆博物馆

图38：苏珊·巴尔

图39：斯托姆·哈尔沃森（Storm Halvorsen），弗雷姆博物馆

图40：苏珊·巴尔

图41：L. 哈克博德（L. Hacquebord），格罗宁根大学

图42：罗伯克鲁克（Robcrook），Wikimedia, https://commons.wikimedia.org/wiki/File:Barentsburg_from_above.jpg

图43：公共领域

图44：中华地图学社绘制

图45：
a. 迈克·皮尔森（Mike Pearson）
b. 迈克·皮尔森

图46：
a. 迈克·皮尔森
b. 苏珊·巴尔

图47：
a. 迈克·皮尔森
b. 迈克·皮尔森

图48：公共领域

图49：
a. 苏珊·巴尔
b. 苏珊·巴尔
c. 苏珊·巴尔
d. 苏珊·巴尔

图50：
a. 苏珊·巴尔
b. 苏珊·巴尔

图51：艾伊尔·克努特（Eigil Knuth）

图52：盖尔·科洛弗（Geir Kløver），弗雷姆博物馆

图53：阿兰·莱特（Alan Light），Wikimedia, https://commons.wikimedia.org/wiki/File:Memorial_Cross_at_Cape_Evans.jpg

图54：苏珊·巴尔

图55：公共领域

图56：
a. 苏珊·巴尔
b. 公共领域

图57：挪威国家图书馆，https://en.wikipedia.org/wiki/Framheim#/media/File:Framheim_med_telt,_hundespann_og_utstyr_rundt_omkring,_1911_(7648958346).jpg

图58：苏珊·巴尔

图59：苏珊·巴尔

图60：
a. 苏珊·巴尔
b. 苏珊·巴尔

图61：苏珊·巴尔

图62：
a. 苏珊·巴尔
b. 苏珊·巴尔

图63：公共领域

图64：苏珊·巴尔

图65：苏珊·巴尔

图66：苏珊·巴尔

图67：苏珊·巴尔

图68：特吕格弗·奥斯（Trygve Aas）

图69：苏珊·巴尔

图70：苏珊·巴尔

图71：苏珊·巴尔

图72：
a. 苏珊·巴尔
b. 苏珊·巴尔

图73：苏珊·巴尔

图74：苏珊·巴尔

图75：
a. 苏珊·巴尔
b. 苏珊·巴尔

图76：苏珊·巴尔

图77：莱斯·洛克图（Lise Loktu），斯瓦尔巴总督版权所有

图78：
a. 苏珊·巴尔
b. 新西兰 Geometria 咨询公司

图79：苏珊·巴尔
图80：特吕格弗·奥斯
图81：挪威海事局
图82：苏珊·巴尔

Polar Science

Polar Cultural Heritage

Too Important to Lose

Susan Barr

Introduction

Fig.1 The polar landscapes are wild and beautiful. Ice and snow are of course characteristic, and Antarctic penguins and Arctic polar bears and walrus are iconic. At the same time both the Arctic and the Antarctic contain exceptional populations of cultural heritage monuments and sites that deepen our understanding of the human interaction with these extreme regions through time. The world would be a poorer place if we were to lose these historical reference points.

▼

The Polar Regions—the High Arctic and the Antarctic—are at the same time both distant concepts for many and at the centre of the current state of global climate change. Even though some still argue about whether the present climate changes are natural or a result of human action, it should be completely obvious to adults all over the world that the climate they have always been used to is no longer the same or as stable as it has been previously in our lifetimes. More than anywhere else in the world it is the Arctic that is demonstrating a warming that is changing the nature and the ecology of this fascinating region. The Arctic is an ice-covered ocean surrounded by land masses, and the last few years have shown an alarming decrease in the extent and thickness of the sea ice. An open Arctic Ocean in its turn creates and strengthens weather patterns that affect the regions to the south. The Antarctic is a huge ice-covered continent surrounded by ocean and the warming effects take longer to become obvious, but they are there, and they too will impact us all.

Already the indigenous peoples of the Arctic are having to adapt to a new way of life that can no longer follow the traditional patterns they have known for generations. Winter and spring hunting on the sea ice is becoming only a memory in areas where there is no longer ice cover, such as the north coast of Alaska. In other areas such as Greenland the ice has become unsafe, so that hunters risk their lives where they before were assured firm and thick ice underfoot. Changes occur too in animal, bird and fish populations, and even modern dwelling places and transport infrastructure have become threatened from effects of the warmer and wetter Arctic climate.

There is much to be said about these two spectacular regions around the Poles. In this book, however, it is the cultural heritage of the polar areas that is in focus. It starts with a discussion of what is meant generally by the term "cultural heritage" before the polar regions specifically are addressed.

Fig.2　Inuit children in Greenland 2018 learn the art of dog driving.

Nature and Culture

▼

Originally there was only nature in the world. Once humans appeared, with their tools and their cultures, they began to modify what until then was pristine nature untouched by humans in any way. The first humans started in the same way as animals before them, by hunting and gathering food to sustain themselves. The effects on nature and ecology were negligible until humans started using their intelligence to moderate the environment around them with sophisticated tools and methods for increasing and rationalising the collection of food for their immediate families, and then their bands and tribes, and ultimately their cities and nations. The dating of the first human tools is still uncertain as scientific investigations continue to uncover evidence that changes our known prehistory. In 2015 finds of primitive tools in Kenya pushed the dating of human tools back from c. 2.5 million years ago to more than 3 million years ago[1].

Our own human species, *Homo sapiens*, is reckoned to have evolved in Africa about 315 000 years ago[2], but the most extreme and profound change in our interaction with the environment occurred around 17 000–12 000 years ago when our ancestors started to alter the natural habitat through the long development of horticulture and agriculture and the reforming of plants and animals to suit our own purposes. Local ecosystems became changed by the burning of vegetation and clearing of land to sow crops. Such practices as these altered the natural environment by opening up areas that had been dominated by dense forest and vegetation to enable some plant and animal species to develop and spread while others retreated or disappeared. Animals were domesticated for easier access to meat, furs and other products, and plant ecology and new soil use changed the natural environment for ever. In exchange for food and security, once wild animals became hunting companions and allies, such as dogs and birds of prey.

Before industrialisation humans still managed to live in a reasonable balance with nature. Gradually the development of permanent settlements led to horticulture, agriculture, animal breeding, fishing and long-term food

Fig.3 A traditional African village in Chad, photographed in 1973.

processing and storage, which enabled larger groups of people to stay in one place all their lives.

Some settlement areas developed into what we call civilisations once they became organised with efficient food production allowing specialisation into activities that could be separated from the daily work of hunting and

Fig.4 Industrial-scale wheat harvesting.

agriculture. Rules and laws for communal behaviour had to be formalised, administrations developed, specialised buildings erected, defence systems organised and written means of maintaining common practices established. The great historical civilisations we know of today were based on agriculture that particularly was situated around large rivers. These waterways provided not only the necessary fresh water for domestic purposes, but were also highways for transport and trade. Their control of water led to a legacy of art and writing that raised them over the previous levels of human settlement and gave them the definition of our first major civilisations. The earliest of these ancient civilisations arose about 5 200 years ago in what is now southern Iraq, in the region of the Euphrates and Tigris Rivers at the meeting place of northeast Africa and southwest Asia. This was followed by development along the Nile River in what is now Egypt. Around 4 500 years ago the same occurred along the banks of the Indus in India, then the Aegean (Mediterranean) culture based on the island of Crete from 4 000 years ago and the Shang Dynasty in China from c. 3 600 years ago[3]—although China already had an unbroken human prehistory of close up to a million years. The northern China area of the Huang He (Yellow River) shows a more continuous history of human development than any other region on earth[4], dating from the so-called "Peking Man" finds of fossilised human teeth and bones of the extinct *Homo erectus* human species at Zhoukoudian near Beijing that date from between 770 000 and 230 000 years ago[5]. The earliest civilisations in the Americas are dated to c. 3 200 years ago in Central America and the Andes.

Despite the existence of these and later population centres around the world it was the advent of industrialisation that really enabled humans to alter the natural environment. Compare the agricultural methods of hand tools, ploughs pulled by animals or people, water carried in bags and buckets and harvesting by hand with the enormous grain fields and mechanised harvesting of today.

Today we alter the course of rivers, erect huge dams to create lakes that flood villages and fields, spray chemicals over natural vegetation, eradicate insects that help to pollinate our crops, lay whole regions under concrete, destroy the rain forests that will never again be able to regenerate from the wasteland they are turned into for growing one-crop types such as soya and

Fig.5
a. Rain forest destroyed for timber in Borneo.
b. Plastic pollution on a beach.

oil palms, genetically modify plants and animals to suit our short-sighted aims, and are on the way to extinguishing the last vestiges of pristine nature. Even the oceans have become a dumping place for our rubbish, pollution and plastics. Is this our ultimate "triumph" of culture over nature?

We will see later how our interaction with nature affects the polar regions.

The meaning or definition of "culture" and cultural heritage

▼

Etymology (the origin and history of words) defines "culture" from two viewpoints: philosophy and logic, and agriculture, hunting and fishing[6]. Regarding the former, culture denotes a society or a group's collected intellectual and material activity (such as Stone Age Culture, Viking Age Culture) or the collective social behaviour, values and norms that pervade in a group. With regard to the latter it is defined as the cultivation of earth, forest, plants or water. This brings us directly back to the contrast between untouched nature and culture, with the latter encompassing human activities and material works including tools, beliefs, the arts and literature.

UNESCO, the United Nations' Educational, Scientific and Cultural Organisation, defines the term "cultural heritage" in the following way:

Tangible cultural heritage which encompasses:
- movable cultural heritage (paintings, sculptures, coins, manuscripts)
- immovable cultural heritage (monuments, archaeological sites, and so on)
- underwater cultural heritage (shipwrecks, underwater ruins and cities)

Intangible cultural heritage covering: oral traditions, performing arts, rituals.[7]

It can be said that everything and anything that has been made or modified by man is cultural heritage, i.e. a cultural object rather than a natural one that can say something about the culture of the person who made it or the situation in which it has arisen and been used. Even a foot trail stamped by the passage of people in the forest or over tundra is evidence of a human action and therefore related to culture and actions. An empty food tin or wine bottle, or a well-used toothbrush can give us clues about types of food of the time, food preservation, drinks and perhaps trade between areas that produce wine and those that do not, oral hygiene practices, and more. Depending on the age of the object and the circumstances we can regard such objects as rubbish

Fig.6 The site of a Swedish scientific expedition at Sorgfjorden, Svalbard that spent the year 1899–1900 here for measurements concerned with the shape of the earth. In the bottom left corner it is possible to make out a rubbish dump of glass bottles. The whole site, including the rubbish dump, is protected by the Norwegian environmental law for Svalbard.

or as interesting objects of study. If the named objects were found at a 100-year old abandoned camp deep in the jungle or in an early 20th century cabin in Antarctica, they would have more historical interest than if they were found in a dustbin in a modern town. So there runs a line between cultural heritage that is considered to be worthy of protection and research and that which is considered to be disposable rubbish. This distinction can be difficult for non-professionals to understand clearly and it is particularly relevant in the polar regions as will be explained further below.

Cultural heritage values

Cultural heritage objects are assigned values that help to define whether they should be considered worthy of protection or not. These values are (unfortunately) not mathematical (although some may be exclusively related to a certain cut-off date) but are dependent on the expertise of the professional heritage workers. The values can also change over time and objects be upgraded or perhaps downgraded. For example, remains from World War II in Norway were originally and understandably cleared away and destroyed in the decades after the war, but have in later years been recognised as heritage worthy of protection however painful the memories may be. Perceptions of heritage values are also different around the world. Where abandoned industrial sites in some areas are considered to be eyesores that must be removed, in other areas they may be protected as local, national or even international heritage. The UNESCO World Heritage List contains examples both of splendid palaces and temples as well as industrial sites.

Cultural heritage objects may be legally protected as single objects such as an individual runestone from the late Viking Age standing alone in a field, or a cache from a historic exploring expedition stored amongst rocks in Antarctica. The tendency nowadays, however, is wherever possible and relevant to include the surroundings that can help to explain the historical

Fig.7 An imperial palace from the Ming and Qing dynasties in China (a) and a coal mining industrial complex in Germany (b), both listed as UNESCO World Heritage.

situation that produced the object. In some cases the object can hardly be understood without the surroundings. An obvious example of this is the historic huts of the Antarctic continent and Peninsula where explorers and scientists wintered under extreme conditions and which, if removed to a museum on another continent, would lose the thought-provoking contrast between a small and fragile wooden building and the overwhelming Antarctic environment. The museum or new situation would have to be innovative and creative to help the visitors to "see and feel" the hut in an Antarctic setting, for example through surrounding films or simulated cold and windy

Fig.8
a. Historic cabins on Cape Adare, Antarctica.
b. The Lingsberg Runestone, Sweden.

/ 119

conditions.

All cultural heritage can therefore provide us with information about the history of our societies and of those that came before. Particularly from the time before written sources the traces of human activities help to bridge the gap between us and our ancestors and they help to explain how we have arrived at our situation today. The development of animal domestication and agriculture through thousands of years can be read from archaeological finds of early tools and all the way down to present-day industrial machines. Where archaeological finds of tools are lacking, rock art and cave paintings can help us to interpret activities of long ago.

Fig.9　Rock paintings from the UNESCO World Heritage Site of Bhim Betka in India. The paintings in the rock shelters here depict some of the earliest examples of scenes of human life, including Stone Age rock paintings which are nearly 9 000 years old.

The Tassili n'Ajjer UNESCO World Heritage site in Algeria contains 15 000 works of rock art dated to up to 12 000 years ago. Amongst many motives they show cattle herding and hunting, as well as wild animals such as antelopes and crocodiles. Even such mundane objects as already-mentioned empty food tins can help to explain history, as exampled by such tins left from an unplanned wintering in 1872–1873 of a boat crew in Svalbard, High Arctic Norway, where new investigations showed that the lead sealing on the cans was an important factor in the death of the men through lead poisoning after eating the contents of the tins.

Fig.10 Empty food tins left at the site where 17 Norwegian hunters wintered in a house in Svalbard 1872–1873. Despite the good accommodation and abundance of conserved food, all the men died. Their deaths have been blamed on scurvy (lack of vitamin C in the diet), but more recent research indicates that lead soldering on the food tins caused lead poisoning that contributed to the men's death.

As our research methods become more and more refined, yet more information can be gleaned from cultural heritage objects both from newer and far older origin. This may also help us with our adaptation for the future, where traditional building methods, for example, are pointing us back to more environmentally-friendly and energy-conserving ways of constructing future housing.

Polar Cultural Heritage

Fig.11 Many hundred simple graves like these can be seen in Svalbard today. These are from $17^{th}-18^{th}$ century whalers who sailed from Europe to hunt whales during the short Arctic summers. Owing to the permafrost, the graves could only be dug very shallow, and the coffins were covered with large stones to keep polar bears and foxes from disturbing the corpses.

Fig.12 Remains of a dwelling site in northeast Canada from the Saqqaq culture, which is particularly associated with early Greenland settlement from Canada about 4 500 years ago. The photograph shows a ring of stones indicating the edges of a skin tent, and a square of stones in the middle where the hearth was placed.

One value of cultural heritage that dominates for most people is not necessarily the value for historical research, but the value of experiencing the object "in person", and particularly when still in situ (= at its original site). A lonely grave in the High Arctic, far from the dead person's home, can create a thread of understanding between the 17th century European hunter who died of scurvy, illness or accident and the modern-day visitor to the site— provided that the latter gives her or himself the time to ponder over events of a few hundred years before rather than only snapping a photograph before hurrying on.

Small objects such as a worn photograph on a hut wall or a broken spade or a stool made of a whale vertebra can set the imagination soaring. An inconspicuous ring of stones on the tundra, perhaps with a few stones in the middle marking a hearth, can give us a hint that an early Inuit family stayed here for a longer or shorter time in an area where we could scarcely imagine

Fig.13　Tourists on a Greek island beach. Large numbers of tourists can create a problem. The famous Italian city of Venice is suffering greatly from its own popularity. 60 000 tourists arrive there each day, and over 20 million during one year. The enormous numbers of tourists who visit the city, famous for its numerous canals and waterways, are driving the Italian population away to more peaceful places to live and are wearing down the most-visited attractions. This and the photograph are just two of many similar examples of historical sites that are being "loved to death" and where the local authorities are considering regulations on the numbers of visitors.

camping even for a short holiday.

Seeing—but not touching!—on a short and organised cruise or bus visit is the closest the large majority of people get to experiencing authentic cultural heritage in its original situation, and of course the experience can help us to understand and hopefully appreciate how and why the objects and sites need to be protected. The experience may be made less awe-inspiring—albeit necessarily so—by signs and barriers that help to shuttle the visitor masses around and through the sites. If we nowadays can be lucky enough to experience walking almost alone on part of China's Great Wall or standing on the Acropolis in Athens on an early winter morning without a tourist group in sight, then we may really be able to communicate with our history in a meaningful way. These are unfortunately rare experiences for most people now.

The Polar Regions

"You know you are in the Arctic if you see a polar bear and in the Antarctic if you see a penguin." This light-hearted way of defining the difference between the two regions is true, but of course only scratches the surface in describing the two areas. Antarctica is the easier of the two to define. Since 23 June 1961 when the Antarctic Treaty of 1 December 1959 entered into force, the Antarctic area has been defined as the land and sea area south of 60° S. Within this area the huge continent of 14 million square kilometres (nearly twice the size of Australia) is about 98% covered by ice that averages 1.6 km in thickness and contains c. 70% of the world's fresh water. The size of the ice-locked area is more or less doubled during winter when the surrounding seas freeze and the ice extends about 1 000 km out from the coast. The main continental mass of Antarctica lies south of 70° S and is usually what is meant when the extreme Antarctic climate is described. However, a land "arm" known as the Antarctic Peninsula, together with associated islands, stretches from the northwest of the continent almost up to the 60° line of latitude. This Peninsula and island area is therefore the easiest to reach by ship and has a relatively milder climate compared with the main continental area, and has thus attracted both sealers, whalers, scientists and tourists in greater numbers than the more-heavily ice-covered continental mass.

There has never been an indigenous population in Antarctica and those who stay there today are either scientists or logistical personnel helping to run the scientific bases. There are no land mammals and very little vegetation apart from lichens and mosses in more-favourable spots. During summer the sea is well populated by a great variety of marine life ranging in size from large whales to krill and micro-organisms, each with their indispensable role to play in the collective ecological system. Seabirds, and not least penguins, are numerous during the summer, but only Emperor penguin males stay on the continent during the winter, where they gather in large groups to incubate a single egg each. Human presence in Antarctica has been ascertained back to at least 1820, when sealers discovered and began to decimate the various

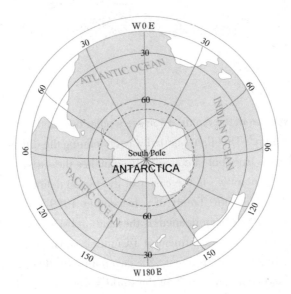

Fig.14 Map of Antarctica.

seal stocks. Although 12 nations claim territory in Antarctica, the 1959 Treaty put these claims on hold and introduced international government by consensus amongst signatory nations. Currently 53 nations have signed the Treaty, the main premise of which is that the continent is only to be used for peaceful purposes and for scientific investigation and cooperation.

The Arctic region is more difficult to define since it is an ocean surrounded by land masses that belong to nations which also extend southwards beyond any definition of "arctic". Five countries have coastlines directly bordering the Arctic Ocean: Norway, Denmark/Greenland, Canada, USA and Russia, while Iceland, Sweden and Finland are included within the definition of land areas lying north of the Arctic Circle of c. 66° N. However, this definition includes areas that otherwise would not be considered to be Arctic, such as the northwest coast of Norway that is tempered by a branch of the warm Gulf Stream. The Research Council of Norway defines the Norwegian Arctic to be only the territories north of the mainland: the archipelago of Svalbard and the volcanic island of Jan Mayen with the surrounding sea areas. This definition is adhered to by this author. The term "High Arctic" is often used to emphasise that it is the northernmost areas of tundra that are meant, where only modest vegetation and no trees and bushes can grow and where the sea is traditionally ice-covered during winter. Unfortunately, climate changes are

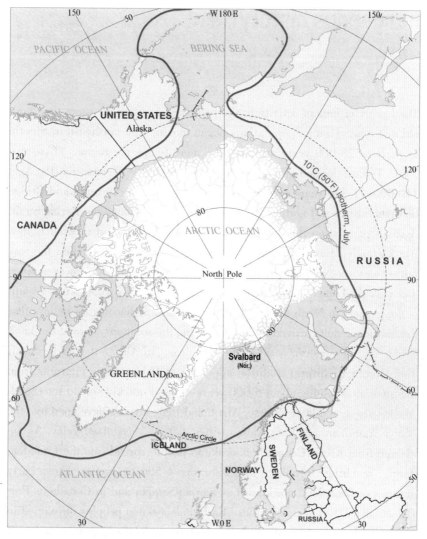

Fig.15 Map of the Arctic. The Arctic Circle is shown in dash lines and the July 10°C mean isotherm in red.

leading to less and less sea ice, even in the High Arctic.

These differences in geographical and political circumstances have led to many contrasts between the two polar regions, but the history and cultural heritage of both regions also have many similarities. These are described below.

Indigenous history

▼

The history of human activity in the polar regions is at the same time both old and new. Human populations began moving into northeastern Siberia as early as 28 000 years ago from the more temperate regions of eastern Asia. Toward the end of the last glacial period of the Pleistocene (Ice Age) 110 000–11 700 years ago, people began crossing the Bering Land Bridge between Siberia and northwest America over what became the Bering Strait. A small population then existed on the Land Bridge and in what is now Alaska (called East Beringia), but a large ice sheet still blocked the way further east into North America until about 12 000 years ago. However, since archaeological sites dating from 14 000–15 000 years ago are known south of the glaciers in North and South America, most archaeologists believe that the first immigrants arrived by boat following the southern shore of the Land Bridge and the western coast of the Americas. In Alaska the Arctic Small Tool (Denbigh culture) tradition appeared as Paleoeskimo migrants from the Siberian Neolithic about 5 000 years ago and quickly spread into South Alaska. These and later Eskimo, Aleut, and Dene arrivals developed into the Native American and First Nation tribes that are recognised today. As ice cleared from Arctic Canada Paleoeskimo people spread into the Canadian Arctic, reaching Greenland and Labrador 4 500 years ago. These early cultures of Greenland are known as Sarqaq/Saqqaq and in Canada as Pre-Dorset. By 2 500 years ago these Early Paleoeskimo peoples developed in place into the Dorset (Tuniit) culture which occupied the Eastern Arctic until the AD 1300 arrival of the Thule culture, a Neoeskimo whale-hunting culture that arose in the Bering Strait region. The Thule culture and people are the ancestors of today's Inuit people of northern Canada and Greenland. These early peoples are known from their archaeological remains.[8] The harsh living conditions meant that their total numbers were never large and no significant settlement groups could develop.

Non-indigenous history

▼

On the other hand, there is the history that encompasses the non-indigenous visitors to the Arctic and those who came from the south and stayed. These were first and foremost the explorers, who were interested both in the geography and in recording the natural conditions, then closely followed by or in conjunction with those who came to exploit and trade the natural resources, be they marine mammals (whales, walrus and other seals, polar bears) and land mammals such as Arctic foxes and reindeer/caribou, or mineral resources such as coal and gold. The exploration of the sailing routes along the northern mainland coasts of North America and Eurasia, known from the Western point of view as the Northwest and Northeast Passages, started with early attempts to find these routes in the 15th and 16th centuries, although Russian settlers and traders had explored parts of the Northeast

Fig.16 The Northern Sea Route near to Cape Chelyuskin, the northernmost point on the Eurasian mainland. This photograph was taken in August 1990 when even icebreakers had problems navigating the Route. Nowadays there is far less ice in the Arctic and the seaway north of Siberian Russia is becoming accessible to many more types of ships.

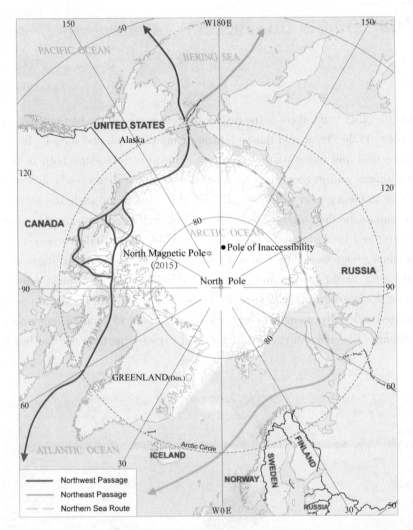

Fig.17 The Northwest Passage (red) north of the North American mainland and the Northeast Passage (green), now known as the Northern Sea Route, north of the Eurasian mainland.

Passage already in the 11th century. Much of the impetus for the historical searches was the hope of finding new routes from Europe to the Orient and the important trade goods that were to be found there, not least silk and spices. Today the route north of Siberia is known as the Northern Sea Route and is again of great interest as a future route between Asia and Europe as the ice conditions along the northern coast of Eurasia change with the warming climate.

East from Asia there was an idea in Europe in the 16th century that a strait, known as the Strait of Anián, separated Asia from the Americas. The strait probably took its name from Ania, a Chinese province mentioned in a 1559 edition of Marco Polo's book. Marco Polo, an Italian trader and explorer, wrote a detailed description of his 24-year long journey to and stay in China in the last part of the 13th century and the Strait of Anian appeared on maps in the 1560s—and in addition in the mentioned edition of Polo's book. The maps portrayed a narrow and crooked Strait of Anián separating Asia from the Americas. The strait grew in European imagination to be an easy sea lane linking Europe with northern China.

The earliest known European exploration of northern coastal North America after the Norse visits to Vinland in the 11th century, was by John Cabot (Giovanni Caboto) in 1497 under the commission of Henry VII of England. From then and up to the first navigation of the Passage in one ship in 1903–1905 by Norwegian Roald Amundsen, the exploration history of the Passage is full of tragedy and heroism. Most famous in this respect was the English expedition in 1845 led by Sir John Franklin when 129 men and two ships disappeared. More than 40 expeditions were sent to search for the lost ships

Fig.18 A famous and iconic painting from 1895 representing the fate of Sir John Franklin's 1845 expedition to find a navigable route through the Northwest Passage north of the North American mainland. His two ships became trapped in the ice off northeast Canada and he and his 128 crew members all perished. The painting "They forged the last links with their lives" by William Thomas Smith is based on some of the evidence that was uncovered about the fate of the expedition.

and men during the following decade, and new details of the story are still being uncovered. What we now know is that the ships became stuck in the ice in the commemoratively-named Franklin Strait on the eastern side of the Passage during the second summer 1846, and all the men perished of scurvy and starvation, many as they attempted to walk southwards to find help. As late as 2014 and 2016 the sunken wrecks of the ships *Erebus* and *Terror* were located near to King William Island east of the Franklin Strait.

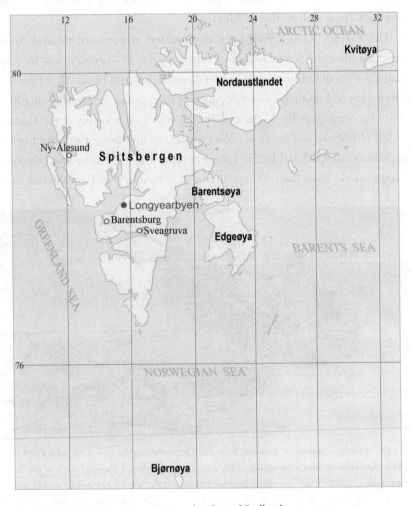

Fig.19 Map of the Norwegian Arctic archipelago of Svalbard.

Svalbard as an example of High Arctic cultural heritage diversity

Svalbard is the name of the Norwegian High Arctic archipelago which lies north of the mainland of northern Norway and stretches almost to 80° N. The whole archipelago is sometimes today mistakenly still called by the historical name Spitsbergen (or Spitzbergen), but this is the name of the largest island only. A faint vestige of the Gulf Stream which brings warm water from the Gulf of Mexico to the west coast of Europe runs up the west side of Svalbard and has throughout history made it somewhat easier to sail farther north in this area than in other parts of the Arctic, although the ice-free season was only 2–3 months a year before the current climate warming period. Svalbard has never had an indigenous population, but since the first time it was without doubt visited—by a Dutch expedition in 1596—its natural resources have been exploited by various waves of temporary visitors.

At the same time as Svalbard was becoming known, European whaling on a commercial scale was growing in importance. The sought-after products

Fig.20 Red Bay in Labrador, Canada, where the extensive remains of a Basque whaling site from the end of the 16th–beginning of the 17th century have been declared a UNESCO World Heritage site. The remains encompass both evidence of the land activities where the oil was boiled out of the whale blubber, a cemetery for 140 whalers, and large and small whaling ships that sank in the bay.

Fig.21 An 18th century engraving showing Dutch whalers hunting Bowhead whales in the Arctic. The mountain appears to be Beerenberg on the Norwegian island of Jan Mayen, where there was considerable Dutch whaling activity in the 17th century.

Fig.22 In 1980 a Dutch archaeological expedition was allowed to open some Dutch whalers' graves in the northwest of Svalbard. The investigations provided a large amount of new information about the whaling period in Svalbard. In agreement with the permission to open the graves they were returned to their original state after samples were taken and it is now not possible to see that they have been disturbed.

were oil and baleen, or whalebone. The oil was used for lighting, lubrication, soap, and in some manufacturing processes for ropes and clothes. The whalebone had such uses as carriage springs, spokes in umbrellas and stiffening in ladies' corsets. European Arctic whaling started in the Labrador-Newfoundland area on the east Canadian coast at the end of the 16th century and spread to Svalbard early in the 17th century. It was particularly from the Basque area in France and Spain around the western Pyrenees and the southern part of the Bay of Biscay that whalers extended their traditional practice across the Atlantic Ocean to the Newfoundland area.

Because of their expertise the Basque whalers were used on the early 17th century British and Dutch whaling expeditions to Svalbard where the Right whale (*Eubalaena glacialis*) was relatively easy to catch. Land stations where the oil could be boiled out of the whale blubber were established on suitable beaches, the most-famous today being Smeerenburg in the northwest of Svalbard. Smeerenburg (Blubber Town) was a Dutch whaling station in use during the first half of the 1600s and the remains of the ovens—where the oil was boiled out—are an attractive tourist destination and also a highly prioritised protected heritage site. Whaling was a dangerous business, especially since the men had to approach the whales in rowing boats in order to get close enough to throw lances. Many hundreds of men in total lost their lives during the summer whaling seasons, either from accidents and drowning or from illnesses. Their graves were shallow as the permafrost

Fig.23 Drawing of a Pomor hunting station in east Svalbard by Norwegian geologist Balthazar Mathias Keilhau who visited the area in 1827. The drawing shows the hunters' log houses and the large orthodox crosses that served both as a religious symbol and as a navigation aid to find the station from the sea. A polar bear is sniffing at a bearskin that is hung over a rack. The mountains portrayed in the background resemble those seen in this area today.

was too hard to dig deeply into, and the coffins were covered by heaps of stones which not least were intended to serve as hindrances to polar bears and foxes which might try to reach the corpses.

The coastal whaling around Svalbard had finished before the end of the 17th century due to depletion of whale stocks, although whaling continued further out to sea (pelagic whaling) without land bases.

Wintering land expeditions took over in Svalbard from the late 1700s by exploiting for their furs the Arctic foxes and polar bears. The first such winterers came from the White Sea area of northwest Russia and were called Pomors after the Russian *nomópbl* (*po*-by and *mare*-sea), which denoted people

Fig.24 A Russian orthodox cross that is one of only a couple still standing in Svalbard. The lower, diagonal arm is missing.

Fig.25 The remains of a Pomor station in east Svalbard. The bottom level of crossed logs shows the dimensions of the huts. The broken red bricks are from the stove that was in a corner of the main hut.

living by the sea. They built small wooden cabins and hunting stations of materials brought with them, sometimes supplemented by driftwood found on the shores, and they erected large wooden orthodox crosses in recognition of their Christian faith.

Almost all the crosses have since disappeared, but the foundations of the hunting stations and huts can still be seen on many of the islands.

A square of wooden logs that are notched together at the corners and a pile of red bricks are one characteristic of these Pomor stations. The bricks were originally used to build ovens for heating the cabins. Graves are also to be found from this period as with all periods of use in Svalbard. Towards the end of the 19th century wintering hunters mainly, but not only, from Norway built simple plank huts around Svalbard and spent one, several or many winters trapping polar bears and Arctic foxes for their valuable skins. Polar bear hunting around the Arctic was forbidden in 1973 by the international Agreement on the Conservation of Polar Bears[9] except for traditional hunting by indigenous people in Greenland, North America and Siberia. Many of these simple huts stand today and a large number are kept in reasonable condition by the cultural heritage office of the Governor of Svalbard (Sysselmann). They are not intended to be used by visitors, but even

Fig.26
a. A typical fox trap where a wooden frame weighted down with rocks is balanced by three pieces of wood, the one stretching under the frame and containing the bait. When a fox puts its head under the frame and grabs the bait, the small pieces of wood come apart and the weighted frame collapses on to the fox and kills it without harming the skin. It is the skin that is the aim of the trapping.
b. A "self-shooting" trap for polar bears. A gun inside the box points out through the opening, with the trigger attached to bait. When the bear puts its head to the opening and grabs the bait, the trigger is pulled and the bear shot in the head. Polar bears have been totally protected in Svalbard since 1973.

so they show us that the trappers right up to the 1970s had very minimum requirements for being able to live and work through the Arctic winters. Some graves from this period too bear witness to the fact that not all of the trappers managed to get through the winter. Some were not well enough equipped and provisioned and died of starvation and scurvy while others died from accidents with small boats or on land. In addition to the huts—which are iconic for Svalbard's history and cultural heritage—remains of fox traps and devices for killing bears can also be seen around the tundra.

Around the year 1900 attempts began at fetching commercially-viable minerals from Svalbard, first and foremost coal, but Svalbard also contains larger or smaller occurrences of iron ore, marble, gold, asbestos, zinc and other minerals. Apart from coal most other minerals are not economically interesting given their High Arctic situation; the logistical challenges are too many. Prospecting and early mining attempts have left behind a highly interesting number of industrial heritage sites, particularly along the west coast of Spitsbergen. Small huts and annexation signs that were erected as

Fig.27 Those hoping for rich rewards from mining minerals in Svalbard brought to the islands machinery, housing for the miners, tools and equipment and left it all behind when the expected profit did not appear. This machinery was brought from England by Ernest Mansfield in the early 1900s to be used to quarry first-class marble, which turned out to be not of saleable quality. Today it stands at the long-since abandoned site of London, or New London as it is also called.

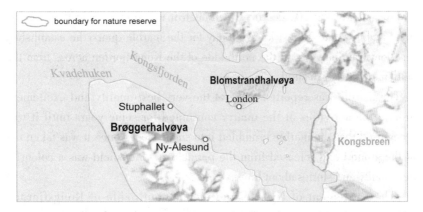

Fig.28 Map section showing the positions of the historic marble quarry London and the present-day research settlement of Ny-Ålesund. This modern map shows an interesting feature of the current climate warming in the Arctic: the name "Blomstrandhalvøya" means The Blomstrand Peninsula, but the rapid melting of the glacier to the north has revealed that the peninsula is in fact an island.

statements of ownership, deeper or shallower tunnels into the hillsides, rail tracks and small wagons for moving the coal or rocks out of the mines, and assorted tools and structures all have both a value for knowledge about this period in Svalbard and also a value for us today to see and gain some idea of the optimism and efforts that some were willing to put into this, usually vain, dream of riches to be scraped and chipped out of the frozen tundra.

One of the hopeful prospectors from this mining period has become a legend in Svalbard history. Ernest Mansfield was an Englishman whose big dream was to find gold. He spent time in the gold areas of New Zealand, Australia and Canada before making his mark in Svalbard during the first

Fig.29 The prospector Ernest Mansfield who established the marble quarry at London (New London) in Svalbard. His big dream was to find gold, but the marble occurrence in Svalbard seemed to be a marvellous investment. Unfortunately the marble quality was not good enough for sale. Mansfield, however, has become a well-known figure in the history of Svalbard.

/ 139

two decades of the 20th century. Although it was finding gold that was his great ambition, he is best known today for the marble quarry he established on Blomstrandhalvøya on the north side of the Kongsfjorden across from the settlement of Ny-Ålesund.

The marble was reported to be of the very best quality and excitement around the activities at the quarry continued for some years until it was recognised that the marble crumbled into small pieces once it was taken out of the ground and released from the permafrost. Mansfield was a colourful personality and stories about him live on[10].

The settlement of Ny-Ålesund on the south side of Kongsfjorden was established in 1916 around a coal-mining claim by the Norwegian company Kings Bay Kul Comp. AS. Coal mining was carried out, with a few interruptions, until a disastrous explosion in the mine in 1962 resulted in it being stopped for good. The settlement has, however, lived on and is today run by the Norwegian authorities as a base for international scientific research. Eleven institutions from ten countries around the world (including China) have established bases there and during the summer field season the place is a hive of international research activity. In addition to this though, Ny-Ålesund contains the largest number of legally-protected cultural heritage buildings (built before 1946)—28—as well as other later buildings with

Fig.30 The settlement of Ny-Ålesund in Svalbard was originally a coal-mining town. Mining finished in 1962 and in later years the Norwegian authorities have invested large resources in developing it as a research centre for international scientists.

high heritage interest. The company Kings Bay therefore works closely with the Norwegian cultural heritage authorities to protect and conserve both the buildings and the settlement itself from detrimental development.

The fact that there was a fully-functioning settlement as far north as 79°N made Ny-Ålesund famous in the 1920s as the starting place for several historic North Pole expeditions. American Richard Byrd, Norwegian Roald Amundsen and Italian Umberto Nobile all led expeditions from the settlement by plane (Amundsen with American Lincoln Ellsworth in 1925, and Byrd in 1926) and by airship (Amundsen with Ellsworth and Nobile in 1926 and Nobile alone in 1928). One of the attractions of Ny-Ålesund today is to see this historic starting place together with memorials to the various expeditions and, not least, the airship mooring mast from 1926 that still stands there on the tundra.

Fig.31 The airship mooring mast in Ny-Ålesund that was erected in 1926 for Norwegian polar explorer Roald Amundsen's expedition to fly the airship *Norge* from Svalbard to Alaska across the North Pole. Italian Umberto Nobile and American Lincoln Ellsworth were jointly responsible for the expedition together with Amundsen. The 16 men in *Norge* became in May 1926 the first people that we know definitely reached the North Pole. The *Norge* landed in Teller, Alaska after 72 hours in the air.

Even though Svalbard would seem to lie too far north to have been affected by World War II it was in fact the scene of several actions that have left their mark today. The Norwegian coal mining settlements of Longyearbyen and Sveagruva, and Russian (then Soviet) Barentsburg were all damaged or destroyed by German ship and submarine attacks. Few reminders of this can be seen today apart from the grave of a Norwegian soldier in the graveyard in Longyearbyen and remains of two German aircraft close to the town. The main cultural heritage objects lie further away from the settlements and include another German aircraft that made a successful emergency landing on the tundra and had to be left there, and the larger or smaller remains of German weather stations. It was important for aircraft and ship movements off north Europe to know as much about the weather as possible and both manned and automatic weather reporting stations were established by German forces for shorter or longer periods in Svalbard as well as across the western Arctic from east Greenland to Franz Josef Land. Remains of several of these can be seen today, the best-preserved being in the north of Svalbard on the Nordaustlandet island. This *Haudegen* station is a highly-prioritised

Fig.32 The German weather station "Haudegen" that was established in northeast Svalbard in September 1944 and operated until September 1945. Haudegen is a unique example of such a war-time Arctic weather station, but is difficult to maintain without altering the building fabric and authenticity of the heritage site. To protect the site from human impact it is forbidden to approach the area around the buildings where numerous remains can be seen on the ground.

cultural heritage site even though the two huts are difficult to maintain without altering the authentic structure too much. The site is very interesting for visitors and because of its fragile nature it is now forbidden to approach too close to the huts and the scattered remains around them. It is, however, easy to view and gain a good impression of the station from points outside the protection boundary.

The newest objects in Svalbard to have been declared protected cultural heritage are larger complexes. In connection with the International Geophysical Year (IGY) 1957–1958 a research station consisting of 10 buildings was established at Kinnvika, Nordaustlandet. The station was used also in 1959 and only sporadically since. It belongs to the Norwegian state

Fig.33 The research station Kinnvika on Nordaustlandet, Svalbard, which was established by a Swedish-Finnish-Swiss expedition for the International Geophysical Year (IGY) 1957–1958. The station has been used only occasionally since. The 10 buildings belong to the Norwegian state and were protected by law in 2009, together with all the contents. Some of the buildings are now infested with mould and fungi, but repairs have been made to a certain extent by the office of the Governor of Svalbard.

and in 2009 the entire station was protected by law. It is possible to visit the complex, although most of the buildings are empty.

In Longyearbyen the aerial transport system for bringing the coal from the mines in Adventdalen to the shipping quay at Hotellneset near to the airport was legally protected in 2003. The system was in use until 1987 and consists of 100 wooden supports for the cables which stretched across 10 km of the tundra, as well as other related structures and, not least, the large spider-like central station that connected the old and the new transport lines in the middle of Longyearbyen.

Fig.34 The central cableway station "Taubanesentralen" in Longyearbyen that joined cables carrying buckets of coal from the mines in three different directions, before the coal buckets were sent on on one cable to the shipping quay. The photograph shows the transport system in use in 1979. In 1987 the last part of the system was closed, and the cables were removed. The wooden pylons that supported the cables across the tundra are still standing and are protected historical monuments.

This great diversity of cultural heritage in Svalbard together with the impressive Arctic nature is the reason why a large part of the archipelago has been suggested by Norway as a possible candidate for UNESCO World Heritage status. Such status can have both positive and negative effects: on the positive side it is intended to promote the protection of the nature and cultural heritage on the basis of their Outstanding Universal Values. On the other hand UNESCO World Heritage status is known to markedly increase the number of tourists visiting such sites or areas and Svalbard today, as a fragile High Arctic region, would already seem to receive more than enough tourism.

Antarctica's cultural heritage

The Antarctic has had no indigenous population nor any mineral prospecting and exploitation, but otherwise has a similar, albeit much shorter, history to the Arctic with visits by whale and seal hunters, explorers and scientists. Sealing by the Antarctic Peninsula started in the early 19th century and soon led to over-exploitation. Whaling followed at the start of the 20th century and land stations for processing the whales were established on Antarctic and sub-Antarctic islands. Once again over-exploitation was a result.

Fig.35 The remains of the early 20th century whaling station at Whalers Bay, Deception Island, Antarctica. Whale processing, when oil was boiled out of the blubber and bones, was carried out on shore here from 1906–1931 and the remains today are from the later period. Deception Island is volcanic and in 1969 the whalers' graveyard as well as buildings and other structures were damaged or destroyed by a volcanic eruption. In 2005 Deception Island was declared an Antarctic Specially Managed Area, ASMA, within the Antarctic Treaty System (ATS) out of consideration for the nature, cultural heritage, research and tourism.

South Georgia

South Georgia is a British Overseas Territory[11] lying in the South Atlantic at 54° –55° S. It thus lies north of the Antarctic region (defined as south of

60°S) but is included here since it was the most important base for Antarctic and South Atlantic whaling during the first half of the 20th century. Sealers had worked in the sheltered bays here as early as 1786 and in 1904 the first of several land stations for processing whales was established in the bay called Grytviken after the Norwegian word for the sealing pots (*gryte*) which were discovered there. Four land stations in particular grew to be large industrial complexes which not only dealt with every piece of the huge whales which were dragged ashore, but also catered in all respects for the men who worked there. In addition to the living barracks, there was a church, cinema, library, sports facilities, canteen and also graveyards associated with the stations. After the end of whaling in the mid-1960s the stations were left to themselves, and both the hard natural conditions and not least ship crews who went ashore to ransack the stations led to severe degradation. The remains

Fig.36 Leith Harbour whaling station on South Georgia which operated in 1909–1965. The large mass of industrial buildings and other structures have remained open to degradation from natural forces and destruction by visitors. The Government of South Georgia has now developed a cultural heritage strategy for the island and the whaling stations which is intended to prevent further damage from human impact. A 200 m protection zone around the station is in force to keep visitors from entering the station area where there is a health danger from asbestos and collapsing structures. Leith and the other historic whaling stations on the island have been thoroughly documented by 3D scanning and photography.

are now recognised as important industrial heritage from the whaling period and are protected from further human interference. Unfortunately they are too dangerous for tourists to visit owing to a large amount of asbestos in the crumbling buildings as well as danger from collapsing structures. The remains have, however, been thoroughly documented by 3D scanning inside and out so that it is possible to make "virtual visits" to the stations (see examples at https://www.youtube.com/results?search_query=geometria+ltd). It is also possible to see the stations from small boats or to walk around the peripheries outside the 200m exclusion zone. The sight of such a quantity of industrial ruins in otherwise pristine sub-Antarctic nature might disturb some people, but the sites are in fact only limited areas that are dwarfed by the magnificent scenery of snow-capped mountains and lush tussock-grass coasts where seals and penguins abound. They are also important sources of historical information both about the whaling methods and how the industry could create such a workplace so far from home, as well as being an experience for visitors to gain an impression of this activity of decades ago on an isolated sub-Antarctic island.

Antarctic history is most famous for the "Heroic Age" of exploration 1897–1922 when such names as Roald Amundsen (Norwegian—in 1911 the first to arrive at the South Pole), Fabian Gottlieb von Bellingshausen (Russian—who circumnavigated the continent in the period 1819–1821), Carsten Borchgrevink (Norwegian/British—in 1899 the first to winter on the continent), Jean-Baptiste Charcot (French—led two exploratory expeditions to Antarctica in the period 1903–1910), Adrien de Gerlache (Belgian—led the first, ship-based expedition to winter in Antarctica in the period 1898–1899), Robert Falcon Scott (British—led the second group to reach the South Pole in 1912), Ernest Shackleton (British—several exploration expeditions between 1901 and 1922) and others added their stories of triumphs and tragedies to the international history books.

In the wake of resource exploiters and explorers, and often together with them, scientists arrived, and it is the scientists who dominate Antarctic activities today. From the International Geophysical Year 1957–1958 and the resulting Antarctic Treaty of 1959 Antarctica has become defined as a continent devoted to peace and science. Currently 70 permanent stations exist across Antarctica, representing 29 countries and every continent. In line with the general expansion of tourism to cover almost every corner of the globe, tourism to Antarctica is also growing and expanding rapidly. Most visitors arrive by ship, the Antarctic Peninsula being the most-visited area both for scientists and tourists, but the Ross Sea area with its collection of Historic Huts is also attractive for tourists, albeit more difficult to reach owing to more severe ice conditions. There are currently also a few possibilities for tourists to fly to Antarctica. Scenic flights in regular aircrafts took place from New Zealand and Australia in 1977–1979, but ceased after an Air New Zealand flight crashed into Mount Erebus in November 1979 with the loss of 257 lives. In 1994 Quantas resumed charter flights from Australia and nowadays some small companies fly limited expeditionary (sports) groups to Antarctica from South America for skiing and climbing, but this is not for the regular tourist. Scientists are flown in to several of the larger bases during the summer season, including the American Amundsen-Scott base at the South Pole. It can be assumed that tourist flights to the continent will increase in the future.

Some iconic polar expeditions

Apart from the Japanese Antarctic Expedition of 1910–1912 there was very little polar activity from Asia in historic times. The Japanese expedition was led by Nobu Shirase, a lieutenant in the Japanese army and with the ship *Kainan Maru*. Despite being relatively unprepared for the conditions they met in Antarctica they reached the Ross Sea where they happened to find the Norwegian expedition led by Roald Amundsen, which had just returned from being the first to stand at the South Pole. A so-called "Dash Patrol" of seven Japanese landed on the Barrier (the Antarctic Ice Shelf) and walked south to 80° 05′ S, at the same time as another group explored the lower slopes of the Alexandra Range on the coast of King Edward VII Land. *Kainan Maru* then returned to Japan. Although a memorial message was left at the furthest-south point, this probably disappeared quickly owing to winds and snowdrift and there are no cultural heritage sites left from this expedition. The expedition is commemorated in Japan at the Shirase Antarctic Expedition Memorial Museum in Akita in northwest Honshu, the main island.

The first buildings on the Antarctic continent

The early 19th century sealers built simple shelters in the Antarctic Peninsula region and remains of these can still be seen. However, the first solid buildings to be erected on the Antarctic continent still stand today—which is a unique record for any continent. In 1898–1900 Norwegian Carsten Borchgrevink led a British expedition which was the first to winter on the continent itself. The site was at Cape Adare at the western entrance to the Ross Sea. Two prefabricated timber huts were erected which had been constructed by a timber firm based just outside Oslo, Norway. The huts were left standing when the expedition departed in their ship *Southern Cross* and they are still standing today as important cultural heritage objects. The New Zealand heritage organisation Antarctic Heritage Trust is working to conserve the huts and their contents of supplies that were left there in 1900,

Fig.37　These two huts at Cape Adare, Antarctica, are the first buildings on the main Antarctic continent. They were erected in 1899 by a British scientific expedition led by Norwegian Carsten Borchgrevink. The log huts were prefabricated in Norway, near to Oslo. The expedition was the first to winter on the continent. The huts and their contents are currently being preserved by the New Zealand Antarctic Heritage Trust, with resources also from Norway and the UK, but the Cape is difficult to get to and stays there can only be short.

Fig.38　The Chinese icebreaker *Xue Long*.

although this is a formidable challenge given the remoteness, extreme climate and logistical problems to overcome. Not least the large numbers of nesting penguins that now cover most of the Cape Adare site must be disturbed as little as possible, even those nesting against the walls of the huts and within the one open hut. With additional resources from Norway and the UK amongst others, a programme of work has begun at Cape Adare which will preserve this unique heritage site for many further years to come. During the Antarctic summer season 2017–2018 the restoration team received support from the Chinese Antarctic research expedition CHINARE-34 which sent 40 persons and its Ka-32 heavy helicopter with the icebreaker R/V *Xue Long* to help with the logistical side of the operation there.

Two other important heritage objects also to be found at Cape Adare are the first grave known to be on the continent and the derelict remains of a hut used by Robert F. Scott's Northern Party, put ashore here from Scott's main base on Ross Island for scientific work during the summer 1911–1912. The grave is the last resting place of the zoologist on Borchgrevink's expedition, Nicolai Hanson, who died of illness during the wintering.

A winter on Franz Josef Land

Polar history is full of amazing stories of survival and ingenuity that still inspire us today. When these stories can be directly associated with a location, a heritage site, remains of boats or shelters that we can stand beside and let our imagination fly back through years or centuries, the reality can affect us deeply and help us to treasure both the stories and the material heritage. One such site is to be found in the middle of the Russian High Arctic archipelago of Franz Josef Land.

Norwegian Fridtjof Nansen led an expedition in 1893–1896 in the specially-designed ship *Fram* to drift in the ice across the Arctic Ocean and perhaps even across the North Pole itself. Nansen's plan was considered by experts of the time to be hazardous to the point of suicidal. There were already by this time numerous examples of wooden ships that had become caught in Arctic sea ice and crushed and a plan to purposely let a ship be frozen into the ice was called madness. Nansen's idea was to prove the existence of an east-west current across the Arctic Ocean—which he did—

and at the same time to investigate whether there could be land in the region around the North Pole, which was completely unknown at the time. The surprising depth of the ocean which was measured during the *Fram*'s drift—almost 3000 m at the most—gave sufficient proof that there could be no land in this area. It was the design of the *Fram* and the massive timber used to build her that allowed her to withstand the enormous ice pressure. The ship can today be inspected inside and out at the Fram Museum in Oslo.

After the first year that proved many of Nansen's theories he became restless and left the ship together with Hjalmar Johansen to strike for the Pole itself with the help of dog sledges. They had to turn back at 86° 14' N and struggled southwards across the shifting and uneven ice to reach land, not knowing that this was Franz Josef Land. Here they constructed what was

Fig.39 The polar ship *Fram* in the museum in Oslo, Norway. The *Fram* was specially designed and constructed for Norwegian scientist Fridtjof Nansen's expedition 1893–1896 to drift over the Arctic Ocean from west of the Bering Strait to north of Norway. This was to prove the existence of a current flowing east-west across the ocean, as well as to carry out many other observations and measurements. *Fram* was then used 1898–1902 on an expedition led by Otto Sverdrup to the unmapped islands off the northeast Canadian mainland, and then in 1910–1914 on Roald Amundsen's expedition to Antarctica when the South Pole was reached for the first time, by dog-sledge. The museum was built specially for the *Fram* and opened by the Norwegian King in 1936. It also houses the smaller ship *Gjøa*, which was the first ship to be sailed all the way through the Northwest Passage on an expedition in 1903–1906 led by Roald Amundsen.

more of a hole in the ground than a hut, where they spent the winter 1895–1896. The winter dwelling was made by scraping a hollow 2 m × 2 m square and a metre deep in the hard ground using a walrus shoulder blade and a ski tip, building up a wall of stones around this a metre high and making a roof of a drift log supporting several walrus skins. The two men lay here in the cold and dark for eight months sharing a worn reindeer-skin sleeping bag for warmth and eating mostly walrus and polar bear meat and fat. Where other winterers under similar conditions often succumbed to starvation and scurvy, Nansen and Johansen managed to keep sane and healthy — in fact increasing their body weight by about 10 kilo ! Once the spring had returned they were able to continue the journey south through the archipelago until they with the largest stroke of luck met up with an English expedition led by Frederick Jackson and were able to return with his ship to Norway. Back in Norway the *Fram* and crew arrived home only a week after Nansen after their drift of

Fig.40 The site on Franz Josef Land where Fridtjof Nansen and Hjalmar Johansen spent the winter in a rough "hut" made by scraping a 2 m × 2 m hollow in the tundra, piling stone walls around 1 m high and creating a roof of a drift log covered with walrus skins. Theirs is an amazing story of how it is possible to survive an Arctic winter with the absolute bare minimum of equipment. The site is now visited by tourists on cruise ships, but it is very vulnerable to human impact on the fragile tundra vegetation around the site and tourism should be strictly controlled.

three years.

In addition to the *Fram* in Oslo, the wintering site on Franz Josef Land can still be seen as a hollow in the ground surrounded by stones and with the drift log in place. Just as the simple huts in the expanse of Antarctica are totally connected with their surroundings, so the remains of Nansen's wintering site are intimately tied to the surrounding nature. Both the hut remains and the nature around them need to be preserved from excess visitation that risks trampling the site and hastening the deterioration.

How the Barents Sea got its name

In 1594, 1595 and 1596 the Dutch navigator and cartographer Willem Barentsz (in English usually written Barents) sailed on expeditions north and eastwards searching for the fabled Northeast Passage to China. Each time he was stopped by sea ice off the northwest Russian coast. On his last voyage in 1596 he first reached an island the expedition named Bear Island (Bjørnøya, the most southerly of the Svalbard islands) and sailed further north to the northwest point of Spitsbergen, the main Svalbard island. Barentsz then returned to Bjørnøya and sailed eastwards to Novaya Zemlja where he and the crew were forced to spend the winter as the sea froze and winter darkness arrived. The 16-man crew built a 7.8 m × 5.5 m house of timber from the ship, calling it Het Behouden Huys (The Saved House). Barentsz died in

Fig.41　The remains of the Dutch winter house on Novaya Zemlya—"het Behouden Huys"—which Willem Barentsz and his crew built for the winter 1596–1597.

June 1597 as the men were making their way southwards in two small boats. Ultimately 12 men survived and were taken back to the Netherlands. The remains of the timber house on Novaya Zemlya still exist and can be clearly seen, but have unfortunately been severely diminished during the centuries

Fig.42   The Russian coal-mining settlement of Barentsburg in Svalbard. According to the Treaty of Spitsbergen (Svalbard) any citizens of any country that has signed the Treaty have the same rights as Norwegians to economic activities in Svalbard. Russia (previously the Soviet Union) is the only country in addition to Norway that has a settlement on the archipelago. The settlement is run on energy from coal and the smoke pollution can be clearly seen in the white landscape.

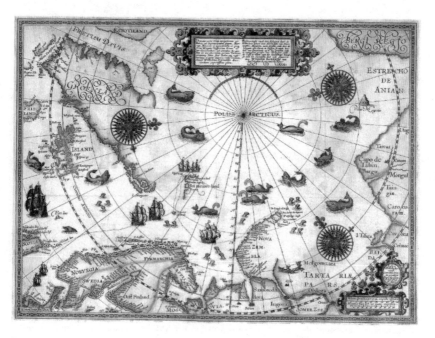

Fig.43   Map of Willem Barentsz' third voyage when a part of Svalbard was discovered and mapped and the voyage ended with an unplanned wintering on Novaya Zemlya.

Fig.44 The three wintering sites of the Swedish Antarctic Expedition 1901–1904: Snow Hill Island, Hope Bay and Paulet Island.

Fig.45 Photos of two of the sites.
a. Remains of the stone hut on Paulet Island where captain C. A. Larsen and 19 men spent an unplanned winter in 1903 after their ship was crushed by the Antarctic ice.
b. Snow Hill. The Swedish expedition house was built in 1902 on a moraine mound composed of matter deposited earlier by ice movement. Unfortunately this has become unstable and attempts have been made to stabilise the edge of the mound.

both by natural causes and by souvenir collectors as well as archaeologists. Nevertheless, it is an important heritage site that should be protected from further degradation by human actions.

The story of Willem Barentsz and his expeditions and death has left us additional intangible cultural heritage through the geographical names that include the Barents Sea, the Russian mining settlement of Barentsburg in Svalbard and the Barents Region of north Norway and northwest Russia.

One expedition, three heritage sites and a sunken ship

In 1901–1904 Swedish geologist Nils Otto Gustaf Nordenskjöld led the Swedish Antarctic Expedition on the ship *Antarctic* captained by Norwegian Carl Anton Larsen. The aim of the expedition was geographical exploration and Nordenskjöld and five men spent the winter in Antarctica as planned in a wooden plank hut the men erected on Snow Hill, a small island off the Antarctic Peninsula. When the *Antarctic* and Larsen returned after the winter to fetch the men at Snow Hill they ran into great difficulties with the ice that prevented passage. Three men were put ashore at Hope Bay with a depot for the men at Snow Hill in case the *Antarctic* could not reach them. The ship battled on but became stuck in the Antarctic ice and had to be abandoned. This was not the only ship to be crushed and sunk by the forces of polar ice, both in the north and south. Captain Larsen brought his crew to safety on Paulet Island where they made a makeshift house of stones. One man of the crew died and was buried in the vicinity. The three at Hope Bay were forced to make their own emergency winter shelter as the *Antarctic* never returned to fetch them. There were now three groups of men in three different places, all waiting through the winter and hoping for some kind of rescue. Amazingly they were all rescued in the end, both by luck and by the appearance of the Argentinian corvette *Uruguay* under Lieutenant Irizar which had been sent to search for the expedition.

From this one expedition alone there still exist the small wooden house on Snow Hill, the remains of the stone hut on Paulet Island together with the grave of crewman Ole Wennersgaard, and the ruins of the emergency hut at Hope Bay, all of which are declared Antarctic protected cultural heritage. And under the sea ice lie the remains of the *Antarctic*.

Antarctic sealers and early Inuit sites

▼

Although not intending at all to imply that sealers in Antarctica in the early 1800s can be compared with the early Inuit in the High Arctic, there are similarities in the remains of dwelling sites of both that enrich our understanding of polar cultural heritage today. Living practices in the Arctic could, of course, vary from region to region, but there is relevance in some comparisons. In both the Antarctic sealing area and amongst early Inuit groups dwellings were established for shorter periods. In the High Arctic early Inuit moved in small family groups according to the availability of food from hunting and fishing. Permanent sites would have tied them to places where the game perhaps moved on and left the people to starve. The housing was therefore simple enough to be erected and used with small means by a mobile family or small group. In winter relatively warm and solid houses could be built in many areas from snow blocks (igloos), while in summer tents or other modest shelters were sufficient. Instead of hauling materials from place to place, as much local material as possible would be used. In the tree-and bushless Arctic stones and earth, animal bones and skins could be utilised. One used what one had for hand. Nor were the dwellings built larger than absolutely necessary to house the family. This was also a way of controlling heating: a simple blubber lamp together with the body heat of people and perhaps dogs were enough when an indoor temperature of few, but sufficient degrees above zero was the norm. When the family moved on a ring of large stones that had held down the skin tent, or a square of stones that had formed the hearth in the centre, would be left behind on the otherwise empty tundra. Coming across these today can firstly challenge the interpretation with the question of how old the site could actually be—stones do not age like wood and vegetation growth might be almost non-existent—but expertise can help to convey the meaning behind the pattern of the stones and give insight into a whole culture that had adapted to an extreme climate.

Sealers in the Antarctic Peninsula area in the early 19th century could spend days or weeks on a freezing, windy shore while harvesting the seals and preparing the skins for transport home. Occasionally they were forced to spend months and perhaps in the worst case a winter on shore. Shelters and dwellings were made, as with the Inuit dwellings described above, from what was at hand, be it a cave or hollow in a rock face, a collection of boulders that could be modified with stones and animal bones such as whale vertebrae, or if they were lucky perhaps some spare timber and sails from the ship. In the c. 150–200 years that have passed the most obvious signs that these were dwellings—tools, utensils, and similar—have mostly disappeared, leaving, as with the Inuit remains, a modest and difficult to understand and interpret site that is far more important for our understanding of Antarctic history than its simple form and materials might imply.

In both these Arctic and Antarctic examples such sites are at risk by being so modest that their importance and origin are not understood. The ring of stones in the Arctic might be rearranged by modern-day campers to suit their own tents and the boulders in the Antarctic sites might be moved or split by geologists seeking the geological history of that region. Information is the key to a good understanding of the values that exist behind even an unusual pattern of stones when it comes to preservation of the most inconspicuous components of our polar cultural heritage.

Fig.46
a. A 19th century sealing site in Antarctica.
b. An early Inuit dwelling site in Arctic Canada.

Common to these sites, and almost all sites in the High Arctic and Antarctica, is the fact that an archaeological excavation often can reveal more evidence and information under the surface, but that there is no, or only very modest, "culture layer depository" in the traditional archaeological meaning

Fig. 47
a. Sealing site at Elephant Point, Antarctica before excavation.
b. After excavation, showing that even here there is a very shallow culture layer.

of chronological layers of different cultures or uses. This is because of the permafrost that prevented digging foundations—and even graves—much below the surface layer, and to the nature of the limited historical use of these areas that did not necessitate building new upon the old as in ancient towns that developed continuously through the centuries. The saying "What you see is what you get" can almost be applied to such sites as shown in Fig. 46 and Fig.47, but even a shallow layer just below the surface can surprise us.

The material evidence of earlier cultures and events

▼

The introductory paragraphs above indicate how the polar areas are full of history. In the Arctic this stretches over thousands of years and has left behind a wealth of cultural heritage sites that are current witnesses to the stories of the past. Despite their often extreme modesty in an overwhelming natural landscape, the sites are as important to the complete history of mankind as are more imposing sites such as the pyramids in north Africa and South America or the Great Wall of China. Without the archaeological sites we would know far less about the spread of mankind from Asia, across the high north of Alaska and Canada, and down the coasts of Greenland. It would be difficult to piece together the history of the earliest peoples who appeared and disappeared as living conditions tipped back and forth from the barely possible to the impossible. It would in addition be difficult for us to imagine and understand how early entrepreneurs scraped their living in a climate that cost hundreds of explorers their lives[12].

Arctic

As indicated above the cultural heritage of the Arctic has broadly speaking two main categories: indigenous heritage and the heritage which has its origins in cultures further south, usually from individuals or smaller groups which moved north mainly to exploit natural resources by hunting, trapping, fishing, whaling and mining, but also for other purposes such as exploration, research and social work—and even war and "cold war". The many-facetted cultural sites and landscapes of the Arctic have values that are important to people, from the individual to the international level. They are our main source of knowledge of how humans interacted with the Arctic nature over time. They reflect the motives behind this interaction and the ways in which the Arctic has been understood and interpreted. They are the inspiration for stories of human endeavours and achievements. For indigenous peoples they are also associated with both the intangible heritage and contemporary living, thus forming a basis for self-definition and sense of place in an historical context that stretches into the future[13].

Internationally significant Arctic sites have qualities that are different from many other sites around the world that are recognised as internationally important. They tend to be less recognisable as physical structures and they challenge the notion of culture as being separate from nature. At the same time they are not hidden by the growth of higher vegetation and by later cultural layers, and the climatic conditions have up to recent times ensured a remarkable preservation of organic materials not seen further south. In addition, the sites that represent the early exploration of the Arctic have gained a mythical quality that has been disseminated in art and literature through many generations[14].

Fig.48 An English painting from 1878 by J. E. Millais entitled "The North-West Passage". The painting expresses the position that the histories of British attempt to map and navigate the Passage had in the imagination of the population. The painting is heavily symbolic, with the old sailor conveying the story of the Passage to his daughter. In the background we see a chart of the north Canadian coast, British naval flags, logbooks from expeditions, a painting of an ice-trapped ship behind the flags and another painting of the English naval hero Lord Nelson.

The High Arctic territories belong to five different nations: Russia, Norway, the Kingdom of Denmark (Greenland), Canada and USA. Each nation has its own laws and policies relating to cultural heritage; a complete inventory of the "population" of cultural heritage sites is therefore difficult to obtain. Fixed cultural heritage sites should reasonably be easier to count than, for example, wandering polar bears or populations of seabirds, but there can be differences of methodology, definition and access to information in each nation that make a total estimate difficult also here. The Environmental Law for the Norwegian archipelago of Svalbard[15] sets 1 January 1946 as the cut-off date for automatic legal protection of all fixed and moveable cultural heritage regardless of provenience and condition. Therefore, there can be protected rubbish dumps from activities during World War Ⅱ or from international scientific activities pre-dating 1946 that have the same level of recognition and protection as the remains of early 17th century

Fig.49 Examples of heritage sites in Svalbard: A hunting cabin (a), mining machinery (b), remains of a whaling oven for boiling oil from the blubber (c) and the foundations of a Pomor hunting station (d).

/ 163

whaling stations or early 19th century hunters' and trappers' simple wintering cabins. This all-encompassing status of automatic legal protection with pre-1946 as the cut-off dating makes as a starting point a potentially uneven definition of cultural heritage in a pan-Arctic connection where other national cultural heritage regimes have their own definitions of cultural heritage worthy of legal protection. In Svalbard a total of 2684 legally-protected heritage sites and monuments has been registered in Askeladden, the national database of protected cultural heritage throughout Norway[16]. This number includes the two younger complexes mentioned above: a large system from the 1950s–1960s for coal transportation from the mines to the shipping quay and a scientific station from the International Geophysical Year 1957–1958 consisting of 10 separate buildings. It does, however, open for the question as to whether a site containing several monuments is to be counted as one or several. As an example, if a 17th century whalers' graveyard is registered as one site, but contains 20 graves, how will the diminishing of the site through coastal erosion—i.e. separate graves being gradually washed into the sea— be registered? By not registering each grave separately it can be difficult to quantify the actual loss.

Fig.50 Trappers' cabins in Northeast Greenland which are very similar to those in Svalbard from the same period, i.e. particularly the 1920s and 1930s.

In contrast the Greenlandic cultural heritage law (*Inatsisartutlov nr. 11 af 19. maj 2010 om fredning og anden kulturarvsbeskyttelse af kulturminder*)[17] sets 1900 as the cut-off date for automatic protection, which excludes the Danish and Norwegian hunter/trapper cabins from the 1920s–1940s that are a large feature of the protected Svalbard heritage. These cabins in both areas were established by the same type of people in the same time period and using similar designs and materials. Happily, however, the cabins in northeast

Greenland are not ignored, but have to a great extent been restored in recent years by a private interest group with the permission of the heritage authorities.

In many regions of the Arctic the component of indigenous heritage is naturally large and can consist not only of the remains of longer—or shorter—term dwelling sites, but also of hunting, burial sites and spiritual practices. These remains can date from as far back in time as several thousand years BC. Specific examples are the alpine ice patch sites in Yukon and Northwest Territories, Canada, which are evidence of caribou hunting that has been radiocarbon dated to more than 9 000 years ago[18], and the caribou-hunting driveline cairns (inuksuk) and tent rings dated to over 4 000 years ago that cover a large area of the Agiak Lake district of Alaska[19]. In north Greenland evidence of the Independence I Paleoeskimo culture exists in archaeological sites dated to 4 200–4 000 years ago[20].

Antarctic

In the Antarctic the earliest sealing sites along the Antarctic Peninsula can, as mentioned above, in many ways be compared to the Paleoeskimo sites in the Arctic. Although dating from as relatively late as the early 19th century, they are modest evidence of the way that men worked and survived through the harsh Antarctic summers while scraping a living from hunting and processing seals. As with the Paleoeskimo sites in the Arctic, these earliest Antarctic sites can easily be overlooked whilst they in fact contain a key to ensuring a complete history of man's activities on the very edge of the possible. At

Fig.51 Paleoeskimo sites from the Independence I culture. They are hard to spot on the bare, stony terrain of northernmost Greenland, but there are the remains of two dwellings in the middle of the land area here.

/ 165

the other end of the scale of Antarctic cultural heritage the wooden cabins of the "heroic" explorers and early scientists are maintained with their contents of bunk beds, tinned foods, dog harnesses, scientific equipment and other artefacts that give an impression to visitors, or to those of us who visit virtually through films and photographs, of the living and working conditions of the intrepid men who in many cases lost fingers and toes, or even their lives, in the search for the ultimate geographical and scientific goals.

A hut is more than just a hut

▼

For many people Antarctic history is primarily understood as the history of the exploration and scientific expeditions of the so-called Heroic Age. This is usually dated from the start of the Belgian Antarctic Expedition in 1897 with the ship *Belgica* under the leadership of Adrien de Gerlache de Gomery, to Ernest Shackleton's death at South Georgia in 1922. The stories of the men and expeditions that occupied this time period have filled hundreds of books already and witness the endurance, initiative, sufferings, deaths, triumphs, good and bad planning and right out good or bad luck that we never tire of reading and hearing about. Several of them have also left tangible heritage on the continent in addition to the intangible heritage that the stories represent. The most iconic and well-known of the material heritage are the "Historic Huts" that are assigned an aura above and apart from any of the later scientific base huts to be found particularly in the Antarctic Peninsula region.

The oldest existing huts on the continent are those established by the expeditions led by Carsten Borchgrevink (Cape Adare 1899–1900), Robert F. Scott's *Discovery* expedition hut at Hut Point (1901–1904), Ernest Shackleton's Nimrod expedition hut at Cape Royds (1907–1909) and Scott's Terra Nova hut at Cape Evans (1910–1913), in addition to Nils Otto Nordenskjöld's hut at Snow Hill (1902–1904) and Douglas Mawson's Australian expedition hut at Cape Denison (1912–1913). All these huts are now defined as important international Antarctic heritage with attempts made to preserve both the huts and their contents as well as possible from the inevitable passage of time and now in later years also from the increasing pressure of tourist visitation. As an attempt to give as many people as possible the experience of this important heritage, Scott's *Terra Nova* hut at Cape Evans has been made available for virtual visiting and viewing through a partnership between the New Zealand Antarctic Heritage Trust and Google[21]. This is a method that many museums, art galleries and important heritage sites are now using to enable more visiting experience without increasing pressure on the sites, and also to allow those who would never get to visit such sites in person to take part in the experience. Although these huts and their contents in themselves are not out of the ordinary, it is of course their siting in Antarctica and the histories connected with each one and each artefact inside the huts that lifts them to the status of particularly iconic international heritage.

The term artefact (or artifact) is used to denote moveable objects that have an historical status, for example through age or association with a recognised event or person. Tools, utensils, clothing or parts of these found during archaeological excavations are called artefacts. Similarly, such objects that are associated with newer events or persons of historical interest are artefacts rather than just objects. Within and around the historical Antarctic huts there is often a wealth of artefacts that were left behind by the expeditions when they returned home. These include food tins, bottles, chemicals and scientific equipment, books and papers, furniture, dog harnesses, packing cases, skins and clothing, kitchen utensils and even dog and penguin carcasses which have become mummified by freeze-drying in the Antarctic climate. Unfortunately, through the years many artefacts have been removed by visitors

Fig.52 Interior of Robert F. Scott's Cape Evans hut. The large number of different artefacts in the Historic Huts, as seen on the photograph here, require skilled conservation in order to keep them from further deterioration. One food tin will, for example, need three different types of conservation: for the paper label, the tin itself and the contents inside.

and others have been handled and moved around within the huts. This can be documented by comparing photographs taken through the years. It is now an illegal act to disturb or take artefacts from the huts except for conservation purposes. Not least the Antarctic Heritage Trust (AHT) of both New Zealand and the UK have carried out large and comprehensive programmes of artefact conservation for some years now. At Scott's Cape Evans hut the New Zealand AHT completed in 2015 a seven-year programme of artefact conservation after more than 11 500 artefacts had been successfully conserved[22].

The Antarctic Historic Sites and Monuments list

As mentioned above the Antarctic is an international area that is not governed by or belonging to one nation alone. Governance is through the Antarctic Treaty System which requires consensus from all Treaty partners. When the Treaty was drawn up in 1959 there were 12 nations involved. Currently there are 53 partners or signatory nations which have acceded to the Treaty and which meet annually to discuss, but at the current time only 29 of these have consultative or voting status which is accorded in relation to a demonstrated and substantial scientific activity, this being the basis for consultative status rather than any territorial claims. In 1968 the Antarctic Treaty Consultative Meeting (ATCM) proposed the compilation of a list of historic sites and monuments in Antarctica, i.e. a list of the most important material cultural heritage on the continent and islands within the Treaty area. The Historic Sites and Monuments List (HSM) has now grown to number 92, although it must be said that from a professional cultural heritage point of view many of the listed monuments and sites would not be considered worthy of such listing were they, for example, to be judged through such a process as that under the UNESCO World Heritage system. The distinction here is the significance related to internationally recognised criteria rather than a more limited national commemoration. Recent discussions within the system indicate that more use may be made in the future of professional cultural heritage expertise in order to raise the status of the Antarctic cultural heritage to become more on par with the professional treatment given to the protection of natural sites and species[23]. Number 87 on the List is a commemorative bronze plaque that marks the location of the first permanently occupied German Antarctic research station "Georg Forster" at the Schirmacher Oasis, Dronning Maud Land. The plaque is well preserved and affixed to a rock wall at the southern edge of the location. The station was occupied from 1976 to 1996 and the entire site was completely cleaned up after the dismantling of the station was successfully terminated on 12 February 1996. Although the fact that the station once existed here can be worthy of commemoration and is an important piece of the history of humans in Antarctica, it can be argued that a monument erected relatively recently to commemorate the fact of a scientific station in itself does not constitute cultural heritage worthy of protection. Such a monument can always be renewed and replaced without any significant loss of heritage value since it was the station establishment that is the heritage "object" and its history will remain even if the plaque should disappear. This is only one of a number of items on the HSM list that heritage professionals would question.

The Historic Huts named above are of course to be found on the HSM list. Sealing sites, which can be considered to be equally important in the history of humans in Antarctica, are not yet there. Ideally the List could be differentiated into monuments and sites that have international importance comparable to listing on the UNESCO World Heritage list and those that have more limited national importance, which is relevant in its own way. Another discussion which has been raised now is whether some heritage objects could be equally well or better preserved by being removed from Antarctica to a museum in their continent of origin where more people would be able to see

Fig.53 HSM17 Memorial Cross at Cape Evans, erected in 1916 to commemorate three British expedition members who died in the vicinity.

Fig.54 HSM76 Ruins of the Base Pedro Aguirre Cerda Station, a Chilean meteorological and volcanological centre situated at Pendulum Cove, Deception Island, Antarctica, that was destroyed by volcanic eruptions in 1967 and 1969.

Fig.55 The tent left at the South Pole by Norwegian Roald Amundsen and his men who were the first to reach the Pole, in December 1911. The tent has been declared an Antarctic Historic Monument (No.80), even though it has long since disappeared under the weight of accumulated snow and ice and the ice movement has transported it away from the original site.

the object and long-term conservation would be easier. Alternatively, some objects could be preserved through film and photo documentation rather than in actual fact. These questions are posed largely from the concern that Antarctica's pristine qualities can be compromised as scientists and tourists encroach further and further into the frozen wilderness and more man-made structures appear.

An interesting example of listing is HSM number 80, the tent erected at 90°S by the Norwegian group of explorers led by Roald Amundsen on their arrival at the South Pole on 14 December 1911 as the first ever to reach this point. The tent is currently buried underneath the snow and ice in the vicinity of the South Pole and has not actually been seen since Robert Scott's group arrived at the Pole a month after Amundsen. This is the only example of intangible heritage on the HSM list, i.e. an object we have only seen in pictures and photographs and which we will never be able to see in real life, but which is strong in many peoples' imagination.

Which monuments and sites have received official HSM status?

To gain a better understanding of the types of cultural heritage that have been endorsed by the Antarctic Treaty System (ATS) as deserving protected status, the list can be found on the ATS website at https://www.ats.aq/documents/recatt/att596_e.pdf . The official Guidelines as to what type of monument or site may be listed give the following criteria for proposals[24]:

    a. a particular event of importance in the history of science or exploration of Antarctica occurred at the place;

    b. a particular association with a person who played an important role in the history of science or exploration in Antarctica;

    c. a particular association with a notable feat of endurance or achievement;

    d. representative of, or forms part of, some wide-ranging activity that has been important in the development and knowledge of Antarctica;

    e. particular technical, historical, cultural or architectural value in its materials, design or method of construction;

    f. the potential, through study, to reveal information or has the potential to educate people about significant human activities in Antarctica;

    g. symbolic or commemorative value for people of many nations.

As with the UNESCO World Heritage List the basic premise is that the proposed monument or site should have significance over and above the local or national; it should be internationally significant with relation to the collective history of Antarctica. The UNESCO World Heritage system uses the term Outstanding Universal Value (OUV). An attempt at analysing the types of heritage on the list is not straightforward as some monuments and sites are complex (i.e. can consist of multiple types) while others are difficult to place without the number of categories becoming too long. The following analysis is therefore only intended to give a general overview, while others may reach a slightly different result.

Commemoration of expeditions	18
Memorials to deceased	12
Commemorating first stations or buildings	11
Associated with the Heroic Age	28
Visit of a Head of State	1
Commemorating a national hero	1
Other memorials	2
Reference for scientific work	11
Whaling	1
Shipwreck	1
Tractor	1

It will be noted that the sum is not 92. Two of the originally listed monuments and sites have later been incorporated into one (numbers 12 and 13 are now together in number 77) while three (25, 31 and 58) have been removed from the list since they no longer exist. It is perhaps in the categories of memorials and commemorations (excluding where graves exist) that questions concerning "outstanding universal value" could be posed. Eighteen listings have been added in the last 15 years (2001–2015). In 2015 it was decided to review the whole system of cultural heritage designation and protection before proposing any more monuments or sites to the list. The review proposal was presented to the Antarctic Treaty System in 2018 and will hopefully improve the processes of cultural heritage management in Antarctica.

Cultural heritage at the Poles themselves?

Of course nowhere is more symbolic of the Arctic and the Antarctic than the North and South Geographic Poles. For centuries these were the greatest goals that explorers could reach, and the attempts cost many lives.

The North Pole is merely a geographical point in the middle of the frozen Arctic Ocean. The ice covering the area around the Pole is constantly shifting, being driven by the winds and currents. The area is too cold and devoid of game for any early peoples to have had any reason to go there and the first people to attempt to reach 90° N were explorers and scientists from nations surrounding the Arctic. Right up to the early 20th century it was uncertain as to whether there were lands or islands at or near to the Pole. Fridtjof Nansen's Fram Expedition in 1893–1896 mentioned above proved by logic based on depth measurements that there could be no land areas lying between the route of the *Fram* from north of eastern Siberia to the Svalbard

Fig.56
a. The site at Ny-Ålesund that was the starting point for the airship "Norge" to fly over the North Pole from Svalbard to Alaska in May 1926. In the foreground is a memorial to the tremendous effort made by the Norwegian carpenter Ferdinand Reinhardt Arild and his men to raise a huge hangar 110 m long, 33 m high and 34 m wide to house the airship prior to leaving. The hangar was built during the winter 1925–1926. Behind the memorial stone one of the fixing points for the stays that anchored down the hangar against the wind can be seen. In the far distance is the mooring mast the airship could use instead of the hangar.
b. The airship *Norge* entering the hangar at Ny-Ålesund in May 1926.

archipelago, i.e. on the Eurasian side. When Norwegian Roald Amundsen together with American Lincoln Ellsworth and Italian Umberto Nobile and crew flew the airship *Norge* in 1926 across the Pole from Svalbard to Alaska, also the area from the Pole to the North American side was observed to have no land amidst the ice. The first people who claimed to have actually reached the North Pole were Americans Frederick Cook in 1908 and Robert E. Peary in 1909, although both these claims have been impossible to prove and have thus been doubted. In 1948 a Russian scientific expedition led by Aleksandr Kuznetsov landed an aircraft at Pole, thus becoming the first undisputed team to stand at the Pole.

The difficulty of proving that one really reached the North Pole lay in the fact mentioned, that this is moving sea ice and not a fixed land point. Erecting a flagpole or a monument would therefore not be of much use. However, in 2007 a Russian expedition with two mini submarines dived to the seabed at 4 300 m and planted a one-metre titanium Russian flag at the Pole point. This is regarded mostly as a stunt, but presumably the flag still stands there and is indeed a piece of cultural heritage, although not representing any territorial claim. With regard to the first expedition that without doubt arrived at the Pole—the airship *Norge*—the starting point in Ny-Ålesund, Svalbard contains cultural heritage from the expedition in the form of a few remains of the huge wood and canvas hangar and the iron mooring tower which were erected during the winter 1925–1926 to receive the airship before it started towards the Pole.

The South Pole would seemingly present an easier placing for a permanent monument, but that is not entirely the case. The 90° S point is covered by an ice layer about 2 700 m thick. With an altitude at the surface of 2 835 m the land surface is actually nearer to sea level. The ice sheet covering is not stationary, but moves at a rate of roughly 10 metres per year in a direction between 37° and 40° west of grid north, down towards the Weddell Sea[25].

Therefore, the position of anything left at the Pole will gradually move away. Roald Amundsen's tent is an illustrative example. Left standing more or less at the Pole itself in December 1911 it was next seen by Robert F. Scott's group a month later. American Richard E. Byrd and Norwegian Bernt Balchen flew over the Pole in November 1929 and did not report seeing the tent. The next to arrive and stand at the Pole was an American group in 1956 who landed an aircraft there. No tent was seen. In fact we can already see on the photographs of the tent that Scott's party took in January 1912 that snow is drifting up around the walls. Some time after this the tent would have been completely covered by drifting snow and gradually pressed down into the ice at the same time as the ice movement carried it away from the Pole point. Attempts to calculate or even physically find the location of the tent today involve many factors but, based on calculations of the rate of movement of the ice and the accumulation of snow, one attempt made in 2010 calculated it probably to lie between 1.8 km and 2.5 km from the Pole at a depth of 17 m below the present surface[26]. Since the tent is now protected from intrusive actions by being accepted to the HSM list in 2005, the actual position has more academic than practical interest.

All structures that are established on the ice in Antarctica as opposed to on firm rock will in the same way as the South Pole tent gradually be covered

Fig.57 Roald Amundsen's wintering base "Framheim" at the Bay of Whales, Antarctica, during his expedition to be the first to reach the South Pole (1911). The wooden cabin was prefabricated in Norway. Together with the many tents and with tunnels and rooms dug under the ice, Framheim became quite a large complex. In the rooms under the ice the men could work through the winter and move from one room to another well protected from the Antarctic cold and storms.

by drifting snow and be pressed down into the ice. Roald Amundsen's wintering base January 1911–February 1912 "Framheim" , by the Bay of Whales, disappeared into the sea some years ago when the ice in which it had become encapsulated broke off as a huge ice floe and drifted out to sea. Similar problems have been faced by several more modern scientific stations, and new stations that are established on the ice now are designed with legs that can be jacked up or the whole station can be moved as necessary.

Robert F. Scott and his four companions who arrived at the South Pole after Amundsen, in January 1912, all perished from cold and starvation on their way back to their base on Ross Island. The bodies of Scott and two of the men were found in their tent eight months later, 18 km from their final depot. The bodies were left in place and the tent was collapsed over them and covered by a high cairn of snow blocks. Again, this would have been covered by drifting snow and buried into the ice in the same way as Amundsen's tent. Today it may only be speculated as to where the grave might currently lie and how deep in the ice. At one point in the future it will perhaps also break off into the sea, frozen into a large piece of the shelf ice.

Do we still have "polar wildernesses"?

The question is most relevant for the High Arctic where the material cultural heritage encompasses evidence of both indigenous and non-indigenous presence all over the area. Indeed, the term "Arctic wilderness" in the popularly-accepted understanding of areas that are untouched by humans, scarcely exists. In the larger areas of the Arctic that have had an indigenous population for thousands of years the tundra can be dotted with stone formations from Paleoeskimo dwelling sites, cairns that point the way along ancient hunting or migration routes, mounds of turf, large bones and stones that indicate a collapsed dwelling, middens (historic "rubbish dumps") of fish, bird and animal bones where a small group of families stayed for a longer time. Very often it takes a trained eye to spot and interpret these historical sites that cannot match the splendour of castles and temples in other areas,

Fig.58   An Inuit "inuksuk" on King William Island, Canada. In the treeless terrain stones could be piled in various ways to leave a message or to point the way.

but which are equally important and irreplaceable for their ability to help us understand and appreciate the past in this region.

In areas with no indigenous population, such as the Norwegian archipelago of Svalbard, humans first began their resource-exploiting activities in the early 17th century, and successive waves of hunters, explorers, prospectors, scientists and tourists have left behind the ruins and relics that we today consider to be heritage worthy of protection as sources of interest, appreciation and, not least, knowledge into the past.

Fig.59 Gåshamna, Svalbard, contains the remains of both early European whaling activities, Pomor (Russian) and Norwegian wintering hunters and a Russian scientific station "Konstantinovka" that was central during a large Swedish-Russian expedition between 1898 and 1902 that was intended to measure the curve of the Earth near to the North Pole region. Fig.6 is from the Swedish station on the same expedition.

In Antarctica there are huge areas of snow and ice landscapes where human beings have never set foot, but which can still have been impacted by air-borne pollutants from populated areas. There are also sites where evidence of human activities and structures have been covered over and "swallowed" by the accumulation of snow. From 1950 to 1960 a mast for meteorological instruments that was established by the Norwegian "Maudheim" expedition of 1949–1952 was covered by snow accumulation to a depth of 8 m so that only the top couple of metres were visible over the pristine snow surface when the site was revisited by the following Norwegian "Norway Station" expedition of 1956–1960. However, in those areas

Fig.60
a. A tourist group being led in a long line across the open terrain in West Greenland.
b. A trail left by the boots of tourists is very visible on the tundra even on stony ground.

where humans can imaginably find reasons and possibilities for visiting and revisiting, and for establishing a presence, there is ample evidence that one can no longer talk of untouched wilderness. In such areas we can see both the acknowledged cultural heritage mentioned earlier that is evidence of historical activities such as exploration and sealing/whaling, as well as the structures belonging to later scientific stations both at the bases themselves and at outlying sites for observation or research. There is also a growing new type of human impact that consists of pathways created by groups of tourists being led to and from bird and animal colonies or cultural heritage sites.

The undeniable attraction of
polar cultural heritage

▼

Even though it is exactly the vision of untouched wilderness that attracts many modern-day tourists to the polar areas, historical sites are in effect often the goals for many of the excursions that visits contain. Of course polar tourists want to see spectacular landscapes and sea ice, polar bears and penguins, tundra and glaciers. But just as often they are taken to see the sites of historical human activity because these are an integrated part of the polar nature we read about and visit today. Nature and culture go hand in hand and it is scarcely possible to see one isolated from the other in the more easily accessible areas.

The archipelago of Svalbard has been rapidly increasing as a cruise destination for many years now. The number of passengers who are put

Fig.61   A group of cruise tourists cluster around a site in East Greenland. This is a natural hot spring site that has been used by the native Greenlanders and which—like cultural heritage sites—can be damaged or degraded by too many visitors.

ashore at sites around the archipelago outside the main settlement areas increased from 29 340 in 1996 to 83 571 in 2017[27]. As mentioned above the coastal areas are dotted with remains of human activities dating from the whalers at the beginning of the 17th century to the German manned and unmanned weather stations from World War II which were established from Bjørnøya in the south to Nordaustlandet in the north. Further inland, in the areas that are not covered by glaciers, there are naturally enough fewer cultural heritage sites, but even so one can here and there find evidence that others have been there before, be it land surveyors' cairns or remains from prospecting or scientific work.

Fig.62
a. The wreck of Roald Amundsen's ship *Maud* lying just below the waterline in Cambridge Bay, north Canada.
b. The wreck was raised by a private initiative and transported to Norway in August 2018. The sturdy dimensions of the hull timbers can be seen here.

What the wealth of websites offering cruises and other tourist visits to the Arctic and Antarctic can tell us is that the polar areas have been opened up to tourism in a way never known before. It is not necessary to delve deep into the climate statistics for most people to have gathered by now that particularly the Arctic is warming rapidly and that there is less sea ice. As a result of this, cruise ships now sail where only ice-strengthened ships previously could go with any degree of safety and success. In 2010 Norwegian polar expeditioner Børge Ousland with three companions sailed a glass-fibre catamaran through both the Northeast and Northwest Passages, thus circumnavigating the Arctic in one season. At the same time the Russian sailing boat *Peter I*, with Captain Gavrilov and crew, also completed the circumnavigation[28]. One hundred years earlier it would have taken the specially-designed and built polar ships such as Fridtjof Nansen's *Fram* and Roald Amundsen's *Maud* years to manage the same. Tour operators that run ships that cruise in the High Arctic during the northern summer turn to the Antarctic for the southern summer season and many tourists that have fallen for the amazing Arctic experience want to go on to visit the spectacular Antarctic.

So now we are getting to the crux of the matter. Climate change and increasing tourism go hand in hand, and with them go the extra impacts on the polar cultural heritage today. Tourism to the Arctic is not new. Gentlemen travellers in their own or hired yachts were sailing to Jan Mayen and Svalbard in the last half of the 19th century both for the travelling experience, for hunting (walrus and reindeer were popular trophies in addition to polar bears) and not least for the collecting of facts about the geography and nature of

Fig.63 An illustration from Lord Dufferin's book "Letters from High Latitudes" which became a popular travelogue from the Arctic after its printing in 1857. The picture shows a glimpse of the top of the volcanic mountain Beerenberg on Jan Mayen, and the ice around the small ship made it extremely difficult for them to get to land on the island. The fact that this was mid-July shows the climatic difference from the mid-1800s to now, when ice like this around the island does not happen any more.

the areas; anything they could record was new information. Lord Dufferin's book "Letters from High Latitudes" describes just one example of such a trip in 1856 and this travelogue achieved great and international success in its time[29].

"Package tourism" for the relatively wealthy without their own yachts rapidly followed, and not least spectacular exploration expeditions such as the Swedish balloon expedition led by S. A. Andrée which attempted to fly from Virgohamna in northwest Svalbard to the North Pole in 1896 and 1897 drew boatloads of tourists to the area both at this time and later. The same Virgohamna was the scene for American journalist Walter Wellman's more or less serious attempts to fly to the North Pole by airship in 1906, 1907 and 1909. Remains of both Andrée's and Wellman's base camps litter the bay today and are since 1974 (Andrée's) and 1992 (Wellman's) protected by the cultural heritage laws for Svalbard.

Fig.64 The remains of two expeditions hoping to reach the North Pole by air now litter Virgohamna in the northwest of Svalbard. Swedish S. A. Andrée made two attempts to travel by hydrogen balloon in 1896 and 1897 and American Walter Wellman tried with an airship between 1906 and 1909. Large amounts of metal filings and scrap were brought to the site for mixing with sulfuric acid to make hydrogen to fill the balloon and airship. This site is one of the most important heritage sites in Svalbard, but it has become disfigured by trails worn by the boots of many visitors. Ideally a boardwalk should be erected here to save further damage.

The fixed and movable objects and artefacts shall neither be disturbed, damaged nor removed. Norwegian Arctic scientist and leader of the Fram Expedition across the Arctic Ocean in 1893–1896, Fridtjof Nansen, visited Virgohamna during a scientific cruise to Svalbard with his own yacht in 1920 and noted (this author's translations)[30]:

> The most of useful and valuable objects, particularly of metal, had by now I presume been plundered, but there was still much left—trappers and tourists had not yet managed to get it all (p.145).
>
> And then the tourists come here and scratch their names everywhere, and help themselves to souvenirs (p.146).

Although it was and is not unusual for private yacht owners to sail to Antarctica, this is a relatively small chapter in polar tourism. In the 1920s it was possible to buy a tourist berth on ships which serviced the whaling and sealing stations on Antarctic islands. The first commercial tourist cruise to Antarctica, however, was organised by the Swedish-American entrepreneur Lars-Eric Lindblad in 1969. His specially-designed ship, *MS Lindblad Explorer*, became famous as the pioneer Antarctic—and also Arctic—cruise

Fig.65 The polar cruise ship *Explorer* pictured in the Arctic in 2006. Although having sailed safely in both the Arctic and Antarctic since 1969, in 2007 she hit hard ice near to the South Shetland Islands in Antarctica and sank. All 91 passengers and 63 crew and guides were evacuated into lifeboats and zodiacs and were extremely lucky that the sea was calm and the weather good as they drifted around for five hours before the first rescue ship could reach them. The ship sank c. 20 hours after hitting the ice.

ship. Unfortunately it also renewed its fame in November 2007 by becoming the first cruise ship to sink in the Antarctic, after hitting hard ice. Happily there was no loss of life.

In the 2017–2018 Antarctic season 43 691 tourists set foot in Antarctica, an increase from almost 37 000 the previous season[31]. Of the 43 691 tourists the two main nationalities were 13 412 from the USA and 7 881 from China. This may not seem many given the size of the continent. It must, however, be borne in mind that the Antarctic summer season is short, the number of sites where it is possible to land tourists are relatively few and therefore relatively well-used, and the visits coincide with the breeding season for Antarctic fauna. As in the Arctic, many boots walking around the same natural or cultural heritage site will unavoidably cause negative impact.

Threats to the Arctic's cultural heritage

Climate change is challenging the preservation of the Arctic cultural heritage as coastal erosion and milder, wetter and wilder weather conditions break down what was once protected by a dry and frozen climate. Work to protect and manage the heritage sites can seem as depressing as the stories of diminishing and threatened polar bear populations.

The long-held axiom of the cultural heritage in the Arctic being "frozen in time" is suffering badly now under the effects of climate change. The axiom became particularly famous in 1987 when a book was published about autopsies that were performed in 1984 and 1986 on the corpses that had been

Fig.66 The graves on Beechey Island, north Canada, of three members of Sir John Franklin's Northwest Passage expedition. The men had died during the expedition's first winter 1845–1946 and the graves were the only evidence of the expedition for many years after the two ships and remaining 126 men disappeared in unknown direction amongst the uncharted islands off the northeast Canadian mainland. The corpses were almost perfectly preserved in the permafrost.

/ 189

buried on Beechey Island during Sir John Franklin's disastrous Northwest Passage expedition in 1845.

One hundred and forty years after the burials it was still possible to recognise the corpses and their clothing and take samples of hair and soft tissues for analysis[32]. Negative effects relating to cultural heritage of the warmer, wilder and wetter Arctic climate are seen through the lack of sea ice causing more coastal erosion, the thawing permafrost that disturbs structure foundations and exposes buried organic material to degradation, more rot and mould destroying wood, more stormy weather that damages fragile structures, and more visitation as mentioned above.

Fig.67 Rot and mould in a building from 1957 in Kinnvika, Svalbard. The building is one of 10 at this historic scientific station that is protected by law. The building has not been used for many years and despite the cold and dry conditions in the northeast of Svalbard the microclimate inside the building has allowed the mould and rot to develop. It is now not advisable for health reasons to enter the building or stay inside for any time. Ideally the building should be preserved through use, but an extensive restoration is needed first.

A map of the 100 most prioritised legally protected cultural heritage sites in Svalbard[33] shows that they without exception are located around the coast. Similarly, this applies to many of the sites all around the Arctic. This was a result of logistical and geographical circumstances: access and appropriate resources were to be found near the coast and people found little reason to

Fig.68 A land-fast ice edge is seen here behind to the left of this author and her daughter. The photograph was taken in 1982 and the lack of sea ice in many of the fjords in Svalbard during the past years has made such ice edges scarcer. They are the vestiges of sea ice cover during winter when the ice against the shore remains longer than the ice on the sea which gets broken up by wind and wave movement. Such ice edges help to protect the shore from wave erosion and their lack now results in increasing erosion along the Arctic coasts.

travel inland. However, as the increasing lack of sea ice, also in winter, removes the barrier against wave erosion that the land-fast ice edge previously could provide throughout much of the summer and certainly the winter, and as wave action itself increases due to more stormy weather in the Arctic, so does the coastal area around the whole Arctic suffer from increased erosion[34].

Thus in turn the coastline moves closer and closer to the cultural heritage sites which ultimately erode into the sea. The erosion can be greatly accelerated in areas with larger ice layers or lenses within the permafrost when the exposed ice thaws and the bonding effect of the ice within the ground sediments is lost, resulting in "slumps" of land. In areas with a large amount of ice-rich permafrost such as the Northwest Territories of Canada slumps can be up to 40 000 m^2 in area with a headwall up to 25 m high[35]. Slumps can severely affect modern as well as traditional settlements and transport routes. Thawing permafrost can in addition add to the stress on cultural heritage by destabilising the foundations of buildings and structures.

Many of the simple, but historically important wooden buildings left by

Fig.69 Two of the many brick buildings in the abandoned Russian mining settlement of Pyramiden in Svalbard. Brick and concrete buildings are not as pliable as wooden buildings, and movement in the thawing permafrost on which the buildings stand causes cracks and destabilisation. Such cracks can be seen around the windows to the left of the entrance stairs. Thawing permafrost is causing extreme problems to infrastructure all around the Arctic as roads, airstrips, pipelines and buildings become destabilised.

Fig.70 One of the oldest buildings in the main Norwegian town of Longyearbyen in Svalbard. Contact with the damp ground has caused the lower part of the wooden walls to rot. The building underwent an extensive restoration in 2007 when this photograph was taken. In many cases such buildings have to be raised up from the ground, which may not be authentic for the particular buildings, but which is a concession to the fact of the milder and wetter climate we now have.

trappers, prospectors, explorers and others in the Arctic were established directly on the frozen ground. As the climate becomes relatively milder and wetter, the wood is exposed to deterioration from rot and mould. This is not necessarily a new situation, but an accelerated one in the new climatic conditions.

Fig.71 The building "Northumberland House" was erected on Beechey Island, Canada, in 1852–1853 during a search organised by the British Admiralty to trace the whereabouts of Sir John Franklin's expedition to find the Northwest Passage. The small building was constructed from masts and other wood salvaged from a wrecked whaler and was built in the vain hope that some of the men from Franklin's expedition might still be alive and be able to return to the Island where it was known that they had spent the winter 1845–1846. The hut was stocked with stores that would help the survivors. Franklin's men were, however, long dead and the contents and materials of the hut were later plundered by others and ravaged by the weather. The pieces of wood and other materials that now lie in and around the hut remains can seem to be only rubbish, but each piece can tell a story of what they once were. It can be difficult for visitors to understand that they should not walk over and through the remains, but each boot print will help to deteriorate the fragile remains even more. This is a common problem for modest heritage sites in the polar areas where the untrained eye does not perceive the historical value of what can seem to be rubbish.

And again, sites and monuments that have rested in peace from visitation through decades and centuries are now increasingly becoming goals for individuals and groups as the barrier the sea ice once presented retreats. Most visitors do of course not intend to have a negative impact, but both the sites and the vegetation and terrain around them are often highly sensitive to even a few boots which can inadvertently dislodge small plants which have protected or stabilised the site, and crush already degrading wooden remains of structures or artefacts. In addition, some few visitors are quite obviously oblivious or indifferent to the damage they do, perhaps by applying graffiti or with careless handling of artefacts or even by taking away "souvenirs" from sites.

Fig.72 a. Worse than the unconscious trampling on heritage sites by visitors is deliberate destruction, often a result of lack of understanding about the site. This graffiti disfigured a highly prioritised 17th–18th century graveyard in northwest Svalbard some years ago. The latest graffiti on a heritage monument in Svalbard occurred as late as 2016 (b) when a tourist sprayed messages on several monuments and other buildings in the main town of Longyearbyen.

Negative effects on cultural heritage help to illustrate climate change

The fact of climate change lies behind much of what has already been written above, and the details of the changing climate have been collected through various natural science disciplines and spread to the general public through the Intergovernmental Panel on Climate Change and many other channels. However, in addition to this extensive work with observations and measurements by the natural science community, the humanities can also inform on and confirm the matter through the disciplines of history, archaeology, historical archaeology and associated work with the material heritage.

History can tell us when a building or structure was first established and perhaps provide details of its situation with regard to the landscape at the time. This in turn may help to document coastal erosion. For example, it might be mentioned in the diary of a scientific expedition member how far the camp was established from the shore, or photographs of a prospecting or mining settlement may show the same. Diaries of others who used the buildings or structures afterwards may also give clues to the rate of erosion. One such example is a trapping station in Svalbard— "Fredheim" —built in 1927 at a safe distance from the shore and consisting of a main house and two smaller buildings. Measurements of the rate of erosion started at the site in 1987, when the main house then stood 17.7 m from the edge. In 2011 the distance had shortened to 8.74 m[36]. Already in 2001, the oldest hut in the complex, which by then lay only 3 m from the erosion edge and was in obvious danger of falling into the sea, was moved 6 m back from the edge. While measurements in 2012 showed that the main house stood 8.5 m from the edge, in 2014 it was only 6 m away. The only alternative to letting the monument go was to move it, and in April 2015 the complete station was actually moved 37 m further in from the shoreline.

Jan Mayen is a Norwegian Arctic island at the northern end of the Mid-Atlantic ridge, the juncture between the continental plates of North America

Fig.73 The trapping station "Fredheim" in Svalbard was established in the 1920s by a Norwegian trapper called Hilmar Nøis who had 35 wintering seasons in Svalbard, 19 at this station, where particularly Arctic fox skins were the target. His history is well known and Fredheim therefore has a special status as a heritage site in Svalbard. The coastline here has always been subject to some erosion, but this increased so much that the entire station was moved in 2015 on to the ridge that can be seen by the flagpole to the left of the photograph.

and Eurasia/Africa. Molten magma continuously wells up in this rift zone and the islands that lie along the ridge are all volcanic. Around the coast of Jan Mayen the sea has a strong erosive force, particularly on the areas with loose volcanic sand and fragmented volcanic debris. Between 1615 and 1645 there was an extensive whaling activity on the beaches where Dutch whalers cut up whale carcasses and boiled the oil out of the blubber. Historical illustrations show broad beaches and several blubber ovens and structures used by the whalers on shore. Through the centuries these beaches have been worn away and there is practically nothing left of the 17th century structures. In 1930 the crew of a Dutch naval vessel placed a 500 kg memorial stone on the inner slope of the beach 80 m from the shoreline. This was documented on film as well as in a written report. In the summer of 2014 the stone stood only 21 m from the sea and the slope was in danger of collapsing together with the stone. The stone was therefore moved to a flatter and safer area further west in the bay. Historical sources therefore prove that the beach was eroded in width almost 60 m between 1930 and 2014. This author first visited the beach in

Fig.74 Kvalrossbukta on Jan Mayen island in 2014. This bay was an important site for Dutch whaling activities in the first half of the 17th century. Many remains of the land station, where oil was boiled out of the whale blubber, were to be seen up to the 1990s, but almost all remains have now been eroded into the sea. The arrow points to a memorial stone which was placed 80 m from the shoreline in 1930, but which had to be moved in 2014 when it was only 21 m from the sea and the slope was in danger of collapsing entirely.

1980 and could then see obvious remains of the innermost structures from the whaling period, all of which have now disappeared into the sea.

In other polar expedition reports and diaries there can be a wealth of information concerning meteorological conditions, sea ice, flora and fauna that can contribute to fill the picture of earlier climatic conditions where there are no long measurement and observation series. The historic state of sea ice in the Arctic has been pieced together with the help of logbooks and diaries from seafarers and whalers; in an article entitled *Piecing together the Arctic's sea ice history back to 1850*[37] Florence Fetterer, principal investigator at the US National Snow and Ice Data Centre (NSIDC), states how sources such as whaling ship logbooks and information on the sea ice edge positions in the North Atlantic between 1850 and 1978 in various sources such as newspapers, ship observations, aircraft observations and diaries have helped to fill gaps and extend the Arctic sea ice record back to 1850.

Insight into permafrost changes have been gained through archaeology. In the same way as the investigations of graves from 1846 on Beechey Island mentioned above, archaeological excavations of 17th century whalers' graves

Fig.75
a. Clothing from the graves of 17th–18th century whalers in northwest Svalbard that were excavated in 1980. Even though the graves were only shallow the permafrost had preserved the clothing and in some cases also skin and hair on the corpses.
b. Excavations in 2016 in the same area showed that the permafrost level had sunk to under the coffins and almost no clothing, skin or hair remained.

Fig.76 Greenland National Museum & Archives in Nuuk that contains collections from all over Greenland and safeguards and promotes the history of Greenland. The collections and exhibitions cover the history from the first culture to arrive in Greenland, the Saqqaq from 4 500 years ago and include the 15th century mummies of six women and two boys who were buried around 1475. Owing to cold and dry conditions both the fully-dressed corpses and the skins they were wrapped in were well preserved when they were discovered in 1972.

in northwest Svalbard carried out in 1980 showed corpses with traces of skin and hair, and with woollen clothes that could almost have been taken off the corpses and put on by the archaeologists. In 2016 and 2017 similar graves in the same area were excavated and such finds were almost non-existent owing to the lowered state of the permafrost that no longer "froze the objects in time".

Similarly, permafrost thawing is destroying organic material in middens in West Greenland that contain evidence of the three main Greenlandic cultures of up to 3 500 years ago—Saqqaq, Dorset and Thule[38]. The realisation that this unique archaeological material can be lost for ever in 80–100 years has prompted targeted research by permafrost scientists in Denmark. Their studies show that the bacteria that normally eat away at organic materials (wood, bone, soft tissues, etc) lie dormant in permafrost, but once that thaws the bacteria become active again and in the process produce heat that in turn helps to thaw more permafrost—an interesting study arising out of an interaction between archaeologists and permafrost scientists.

Using technology to help protect cultural heritage

Repair and restoration are traditional methods of protecting and prolonging the life of buildings and structures. In seldom cases actual moving of a monument such as a small building threatened by erosion has also been used, as mentioned above concerning the trapping station "Fredheim" in Svalbard. Monitoring the effects of natural impacts such as erosion and degradation of wooden materials, and of human-caused impacts such as wear and tear on the heritage sites and surrounding vegetation, is an important method and such work will continue.

However, in addition to the traditional methods attention is increasingly being paid to the use of new technology in these respects. Drones can be used to monitor sites and measure changes such as erosion increase without the operator having to set her own boots on the site. Amongst many examples are the drone monitoring of the rapidly increasing erosion along the coast of the Canadian North-West Territories[39], and site surveys of heritage sites in northwest Svalbard by drones.

Fig.77 A composite photograph taken from a drone showing a 17th–18th century whalers' grave site in northwest Svalbard. The sea and beach are at the bottom of the picture. The many small oblong shapes in the middle are the separate graves which were covered with rocks to protect the corpses from polar bears and foxes. The light-coloured area around and amongst the graves is the result of visitors' boots wearing away the scanty tundra vegetation. This demonstrates why some heritage sites should have restrictions limiting tourism inside the main site area.

In addition, the development of monitoring satellites that cover the Arctic area opens an exciting field of possibilities for remote information gathering. The European Union *Copernicus* Programme is interesting in this respect. It is aimed at developing European information services based on satellite earth observation and in situ (non-space) data[40]. The introduction of remote-sensing tools opens a whole new world of cultural heritage monitoring in remote environments and gives the opportunity for far more intensive studies of particular sites without the detrimental accompaniments that traditional expeditions to the areas unavoidably give, including air and sea transport emissions and direct human impact on the sites.

A further technological advancement that has been introduced to and embraced by heritage professionals is the use of scanning technology. Detailed measurements, photographs, scaled drawings and written descriptions have traditionally been the staple methods of documentation of monuments and sites. To enable this documentation to speak for itself, independent of the actual object in question (for example if the object should later be lost through erosion, fire, or natural degradation), extreme care and accuracy are required which in turn means time and other resources spent

Fig.78
a. One of the large industrial complexes left after 20th century whaling activity on South Georgia, in the sub-Antarctic. These are important cultural heritage sites, but are not possible to preserve in all their complexity.
b. The British South Georgia authorities, together with Norwegian support, have therefore had comprehensive and detailed 3D scanning documentation made of the entire insides and outsides of the four main stations. This illustration could seem to be an aerial photograph, but is a scan composed of the raw material obtained on the ground. Through this documentation the stations can now be virtually visited either by studying individual scans or by the animated scans that allow the virtual visitor to "walk through" the sites and the buildings.

Fig.79 The abandoned whaling stations on South Georgia are riddled with asbestos. It is therefore forbidden for health reasons to approach nearer to the stations than 200 m. In 2015 this author participated in a survey of the stations together with the British authorities. It was mandatory to use full protective covering against asbestos dust. The helmet and face mask were not very comfortable to use over time!

in the field in gathering the documentation. By using 3D laser scanning, extremely complicated heritage sites can be captured in a short time by a pair of operators. Work to then sort and present the data after collection admittedly takes time, expertise and appropriate softwear and computer capacity, but this work is done back in the office and the actual field time is short and effective. This author has been involved in the total scanning of the complicated industrial and now deserted whaling stations on South Georgia in sub-Antarctica where two operators have used only a few days in the field to cover an entire station inside and out.

As mentioned above further examples of the results can be found at https://www.youtube.com/results?search_query=geometria+ltd) and the documentation secured in this way is intended to be used not only for virtual visits and tours of the historical whaling stations, but also for a variety of research projects concerned, for example, with station layout and architecture, land use, more general whaling history and for examining details of buildings and structures, perhaps with regard to possible protection or restoration of specific elements.

The technique is used in the High Arctic as well. In 2010 a laser scan was made of the historical site of Fort Conger at Lady Franklin Bay, Ellesmere Island, Canada. The paper written about the project explains that:

Fort Conger is currently at risk because of the effects of climate change, weather, wildlife, and human activity. In this paper, we show how 3D laser scanning was used to record cultural features rapidly and accurately despite the harsh conditions present at the site. We discuss how the future impacts of natural processes and human activities can be managed using 3D scanning data as a baseline, how conservation and restoration work can be planned from the resulting models, and how 3D models created from laser scanning data can be used to excite public interest in cultural stewardship and Arctic history.[41]

The paper gives an excellent description of the use of this technology, which can be applied to all sizes and types of objects and sites.

How to lessen the impact of tourism?

The fact that tourism to the High Arctic and Antarctica continues to increase, leads heritage managers to act not only by introducing regulations and limitations, but to a large degree also by presenting the visitors and the tourism operators with as much information about the various historical sites as possible. Once a visitor is told or can read that this or that site was actually the very place where an important historical event took place, or is an amazingly preserved example of the will and the way to survive under far more severe climatic conditions than one meets today, then in almost all cases he/she will treat the sites with reverence and care, taking only away some photographs and a memory of a unique experience relating our own time to events long past.

In this situation of need-to-know and need-to-inform, the historical information around the various monuments and sites in the polar regions continues to grow and in turn provides material for more popular books about the history of these regions which hopefully in their turn increase serious interest in the regions. Very much is, however, up to the operators, crew and guides on cruise ships to the polar regions. It is their responsibility both to have a clear understanding not only of the relevant laws and regulations, but also of the ethical dimensions of intrusion on the sites and the environmentally most-protective ways of visiting fragile areas. In turn they can then inform and carefully guide the visitors. It should not least be clearly understood that some areas and sites should just not be visited at all, as any kind of human impact can have negative results. This can include, for example, increasing erosion or wear and damage to stabilising vegetation or scattered historical objects.

Many of the operators who run tourist cruises to the Antarctic are members of IAATO, the International Association of Antarctica Tour Operators (https://iaato.org/home). The corresponding organisation in the Arctic is AECO, the Association of Arctic Expedition Cruise Operators (https://www.aeco.no/). In addition to being member organisations for tourism operators, the two organisations promote responsible polar tourism by developing guidelines both to inform operators as to best practice and to mitigate as far as possible the negative effects of visitation on sites of cultural and natural heritage, wildlife and local settlements or scientific stations. This is an essential move within the business and is greatly welcomed by area management authorities. Unfortunately, not all commercial operators are members and private expeditions are certainly not.

In response to an argument often raised by tour operators and others, that cultural heritage sites are the heritage of all of us and therefore should be open to visitation by all, it can be countered that internationally important heritage sites all around the world are currently being considered with respect to limitation of or complete closure to visitation. The UNESCO World Heritage sites of the Chilean Easter Island statues, the Egyptian pharaohs' tombs in the Valley of Kings, and the Machu Picchu Inca city in Peru are examples of sites where excessive tourism is damaging or destroying to such

Fig.80 The Hal Saflieni Hypogeum UNESCO World Heritage site in Malta is a complex of subterranean stairs and chambers that has survived for over 5 000 years. The deteriorating effects of humidity from visitors have led to the authorities strictly controlling and limiting access to the site.

an extent the details that have made them famous that visitor limitation is being or has been imposed. The Lascaux Cave in France, which contains some of the most amazing Palaeolithic cave paintings ever found—dating from up to 20 000 years ago—was closed to the public in 1963 when it was realised how much damage the visitation was causing through body heat, humidity and carbon dioxide emissions. Twenty years later a "true copy" was opened nearby to give some satisfaction to the public. Today much could also be achieved through virtual reality experiences. The exceptional tomb of the boy pharaoh Tutankhamun in Egypt's Valley of the Kings was closed for restoration between 2010 and 2014 owing to damage from the millions who have visited since its discovery in 1922. In May 2014 a full re-creation of the chamber in exact detail was opened for the public. Although the original has since been reopened, closing it for good must be considered. Closing or limitation of access to certain particularly important polar heritage sites would therefore not be exceptional, but on the contrary would have a firm basis in current international cultural heritage management. In fact this has already been practised in Svalbard where nine particularly important heritage sites received access limitations in 2009.

Thoughts about the future

▼

Having worked exclusively in and with the polar regions for the past 40 years (from 1979) it is easy for this author to testify to the changes that have occurred already. Where sea ice and snow on land earlier limited the season for field work in Svalbard to 2.5–3 months each summer and the few cruise ships that gradually appeared over the years only ran trips in July and August, it is now not unusual to have short cruises starting up at the end of March and the season ending in October, i.e. it is the length of daylight that now limits such tourism rather than any sea ice hindrances.

The polar regions are spectacular, and it is very understandable that people wish to visit them. Some impetus is also to be found in the constant news of shrinking areas of snow and ice and alarming warming of the Arctic and the Antarctic ice rim. A feeling spreads that one must experience the polar world before it disappears, and in this way all our travelling helps to add to the impact on our climate. It is said that visits to the regions are so moving that they create ambassadors out of the tourists who will spread the word about the need to preserve the region. However, there is a fine line between bringing more and more people to experience and perhaps become ambassadors for preservation and actually preserving the areas by not physically impacting them. The marketing idea of recent years that promotes "eco-friendly", ecologically-positive cruises to "see" climate change in the Arctic and observe polar bears and seals that are suffering from the changes, ignores the fact that one cruise ship spews out as much CO_2 as 13 000 diesel cars per day in addition to extensive atmospheric particles and NO_x pollution[42]. Less visible is the effect that engine noise has on marine life and, in areas that do still have sea-ice cover, the challenges caused by ships breaking their way through the ice and opening channels that can both disturb wildlife and break up the white surface of the sea ice to give dark surfaces that absorb more of the sun's heat, thus creating an additional warming effect. Undoubtedly cruise ship owners can already see "the writing on the wall" with regard to the need to redesign ships to meet and mitigate

Fig.81 The Geirangerfjord in Norway is a UNESCO World Heritage site, with its steep mountains enclosing the fjord on both sides. The beauty of the scenery here makes it a favourite place for cruise ships to visit and there can be several large ships in the fjord at one time. The pollution from the ships' fuel is easy to see and is spoiling both the scenic effect and the health of the area and the people who live there.

these negative effects. This will take time, but hopefully new regulations and improvements concerning fuel types and pollution control in new cruise ships will to some extent gradually relieve the situation.

In my opinion we must accept that large areas of the High Arctic and the Antarctic should be declared off-limits for group tourism, some areas for any tourism at all, while others may be visited by individuals or small groups that follow strict guidelines for causing minimal impact. It seems to be more easily accepted that such strict measures are put in force to protect bird or animal species so that they can continue to exist and propagate, while unique heritage sites that are non-reproducible are inclined to be considered freely available by some sort of right for everyone. As early as in 1973 the islands known as Kong Karls Land in eastern Svalbard were closed for all visitation all year round except for occasional scientific visits by permit. This was to protect the denning area of polar bears. The visiting ban has never been seriously questioned since we can all understand the need to protect polar bears. Likewise the small 2 km × 3 km low island of Moffen north of Spitsbergen has been declared a nature reserve in order to protect the walrus

Fig.82 Garðar in west Greenland was the seat of the bishop of the Norse settlements in Greenland in the $12^{th}-14^{th}$ centuries. The site is beside the current settlement of Igaliku. Many large stones from the Norse church and bishop's residence can still be seen in situ, while others have been incorporated into the new buildings. The site is in breath-taking scenery in a relatively fertile part of Greenland.

which haul out and rest there. Strangely enough there is far more opposition to similar closing of areas for their cultural heritage importance. We may need to return to the mindset of several decades ago when it was accepted that amazing television and film documentaries were our only possibility of seeing deep into the jungles, rain forests, extensive deserts and ice-locked Antarctica—places where we would never set foot in actual fact. Virtual reality technology can perhaps help us to save the cultural and natural heritage of our vitally important polar areas until such a time as technology also helps us to visit the places in actual reality with no negative impact on the surroundings at all.

Notes

1. https://www.britannica.com/event/Paleolithic-Period
2. https://www.britannica.com/science/human-evolution
3. http://www.historyworld.net/wrldhis/PlainTextHistories.asp?historyid=ab25
4. http://www.historyworld.net/wrldhis/PlainTextHistories.asp?gtrack=pthc&ParagraphID=bei#bei
5. https://www.britannica.com/topic/Peking-man
6. de Caprona, Yann 2013: Norsk etymologisk ordbok. Kagge forlag AS, Oslo. Quotes are translated here by this author.
7. http://www.unesco.org/new/en/culture/themes/illicit-trafficking-of-cultural-property/unesco-database-of-national-cultural-heritage-laws/frequently-asked-questions/definition-of-the-cultural-heritage/
8. I am grateful to Director William Fitzhugh of the Smithsonian Arctic Studies Center in Washington, USA for his input to this paragraph.
9. http://pbsg.npolar.no/en/agreements/agreement1973.html
10. Susan Barr, David Newman and Greg Nesteroff 2012: Ernest Mansfield, Gold or I'm a Dutchman. Biography. Akademika Publishing, Trondheim, Norway.
11. http://www.gov.gs/
12. Barr et al. 2013: Assessment of Cultural Heritage Monuments and Sites in the Arctic. Arctic Council (SDWG) Project #P114:2
13. Ibid.:6, 7
14. Ibid.:7
15. https://lovdata.no/dokument/NL/lov/2001-06-15-79?q=svalbard
16. https://www.riksantikvaren.no/Veiledning/Data-og-tjenester/Askeladdenand Kulturminneplan for Svalbard 2013-2023, Sysselmannen på Svalbard Rapportserie Nr.1/2013, p.31.
17. http://dqbglhnbfffy1.cloudfront.net/fileadmin/user_upload/Inatsisartutlov_nr_11_af_19_maj_2010_om_fredning_og_anden_kulturarvsbeskyttelse_af_kulturminder.pdf
18. http://pubs.aina.ucalgary.ca/arctic/Arctic57-3-260.pdf
    http://ipy.nwtresearch.com/Documents/Hunters%20of%20the%20Alpine%20Ice.pdf
19. https://www.nps.gov/gaar/learn/historyculture/landscape-archaeology-at-agiak-lake.htm
20. https://web.archive.org/web/20090113130338/http://www.natmus.dk/sw18660.asp
21. https://www.nzaht.org/explorer-bases/scotts-hut-cape-evans
22. https://www.nzaht.org/pages/history-of-the-projects
23. Read more in https://www.tandfonline.com/eprint/KnEmZnK4YazRB9b4vXS7/full

24. https://www.ats.aq/documents/cep/Guidelines_HSM_V2_2009_e.pdf
25. https://en.wikipedia.org/wiki/South_Pole#Ceremonial_South_Pole
26. Polar Record / Volume 47 / Issue 03 / July 2011
27. http://www.mosj.no/no/pavirkning/ferdsel/cruiseturisme.html
28. https://www.seilmagasinet.no/innhold/?article_id=31015 and http://www.yachtingworld.com/news/world-record-for-russian-crew-in-arctic-7471
29. Dufferin, F.H.T.B. 1857: Letters from High Latitudes. John Murray, London.
30. Nansen, Fridtjof 1920: En Ferd til Spitsbergen. Kristiania.
31. Figures from IAATO, the International Association of Antarctica Tour Operators.
32. Beattie, Owen & Geiger, John (1987). Frozen in Time: Unlocking the Secrets of the Franklin Expedition. Saskatoon: Western Producer Prairie Books. ISBN 0–88833–303–X.
33. https://www.sysselmannen.no/globalassets/sysselmannen-dokument/trykksaker/katalog_prioriterte_kulturminner_paa_svalbard_versjon_1_1_2013_komprimert.pdf , page 8
34. https://www.livescience.com/13746-arctic-coast-erosion-climate-change-ice.html
35. http://www.nwtgeoscience.ca/project/summary/permafrost-thaw-slumps
36. Note 61, p. 71 in Kulturminneplan for Svalbard 2013-2023, Sysselmannen på Svalbard Rapportserie Nr.1/2013.
37. https://www.carbonbrief.org/guest-post-piecing-together-arctic-sea-icevhistory-1850
38. http://sciencenordic.com/climate-change-destroying-greenland%E2%80%99s-earliest-history
39. http://www.cbc.ca/news/canada/north/drones-monitor-nwt-arctic-shoreline-erosion-1.3897042
40. http://www.copernicus.eu/main/overview
41. http://polarheritage.com/content/library/Arctic_Application_of_3D_Laser_Scanning.pdf
42. Reported in the newspaper Klassekampen 19 July 2018.

Picture Sources

Fig.1: Susan Barr
Fig.2: Susan Barr
Fig.3: Richard Barr
Fig.4: Joe Donovan, https://commons.wikimedia.org/wiki/File:Harvesting_the_wheat_-_geograph.org.uk_-_144209.jpg
Fig.5:
a. Rhett A. Butler/ Mongabay.com
b. Vaidehi Shah, https://commons.wikimedia.org/wiki/File:Litter_on_Singapore%27s_East_Coast_Park.jpg
Fig.6: Susan Barr
Fig.7:
a. Wikipedia
b. Spyrosdrakopoulos, Wikimedia, https://commons.wikimedia.org/wiki/Category:Zollverein_Coal_Mine_industrial_complex#/media/File:1417_zeche_zollverein.JPG
Fig.8:
a. Marcus Thomassen/Fram Museum
b. Berig, https://commons.wikimedia.org/wiki/File:U_240,_Lingsberg.JPG
Fig.9: Susan Barr
Fig.10: Susan Barr
Fig.11: Susan Barr
Fig.12: Susan Barr
Fig.13: Mstyslav Chernov, Wikimedia, https://commons.wikimedia.org/w/index.php?search=Tourists+Mykonos+Island+Greece&title=Special%3ASearch&go=Go&ns0=1&ns6=1&ns12=1&ns14=1&ns100=1&ns106=1#/media/File:Little_Venice_quay_flooded_with_tourists._Mykonos_island._Cyclades,_Agean_Sea,_Greece.jpg
Fig.14: Illustration from Wikipedia, redrawn by Chinese Map Publishing House
Fig.15: Illustration from Wikipedia, redrawn by Chinese Map Publishing House
Fig.16: Susan Barr
Fig.17: Illustration from Wikipedia, redrawn by Chinese Map Publishing House
Fig.18: Public Domain
Fig.19: Wikipedia, redrawn by Chinese Map Publishing House
Fig.20: Susan Barr
Fig.21: Public Domain
Fig.22: Susan Barr
Fig.23: Public Domain
Fig.24: Susan Barr
Fig.25: Susan Barr
Fig.26:
a. Susan Barr
b. ØysteinWiig
Fig.27: Susan Barr
Fig.28: © Norsk Polarinstitutt, npolar.no. Redrawn by Chinese Map Publishing House
Fig.29: *Ernest Mansfield, Gold or I'm a Dutchman* by Susan Barr, David Newman, Greg Nesterorff, 2012
Fig.30: Susan Barr
Fig.31: Susan Barr
Fig.32: Susan Barr
Fig.33: Susan Barr
Fig.34: Susan Barr
Fig.35: Susan Barr
Fig.36: Susan Barr
Fig.37: Marcus Thomassen/Fram Museum
Fig.38: Susan Barr
Fig.39: Marcus Thomassen/Fram Museum

Fig.40: Susan Barr
Fig.41: L. Hacquebord / University of Groningen
Fig.42: Robcrook, Wikimedia, https://commons.wikimedia.org/wiki/File:Barentsburg_from_above.jpg
Fig.43: Public Domain
Fig.44: Illustration: Susan Barr, with map base from Googlemaps
Fig.45:
a. Mike Pearson
b. Mike Pearson
Fig.46:
a. Mike Pearson
b. Susan Barr
Fig.47:
a. Mike Pearson
b. Mike Pearson
Fig.48: Public Domain
Fig.49:
a. Susan Barr
b. Susan Barr
c. Susan Barr
d. Susan Barr
Fig.50:
a. Susan Barr
b. Susan Barr
Fig.51: Eigil Knuth
Fig.52: Geir Kløver/Fram Museum
Fig.53: Alan Light,Wikimedia, https://commons.wikimedia.org/wiki/File:Memorial_Cross_at_Cape_Evans.jpg
Fig.54: Susan Barr
Fig.55: Public Domain
Fig.56:
a. Susan Barr
b. Public Domain
Fig.57: National Library of Norway, https://en.wikipedia.org/wiki/Framheim#/media/File:Framheim_med_telt,_hundespann_og_utstyr_rundt_omkring,_1911_(7648958346).jpg
Fig.58: Susan Barr
Fig.59: Susan Barr
Fig.60:
a. Susan Barr
b. Susan Barr
Fig.61: Susan Barr
Fig.62:
a. Susan Barr
b. Susan Barr
Fig.63: Public Domain
Fig.64: Susan Barr
Fig.65: Susan Barr
Fig.66: Susan Barr
Fig.67: Susan Barr
Fig.68: Trygve Aas
Fig.69: Susan Barr
Fig.70: Susan Barr
Fig.71: Susan Barr
Fig.72:
a. Susan Barr
b. Susan Barr
Fig.73: Susan Barr
Fig.74: Susan Barr
Fig.75:
a. Susan Barr
b. Susan Barr
Fig.76: Susan Barr
Fig.77: Lise Loktu © Sysselmannen på Svalbard
Fig.78:
a. Susan Barr
b. Illustration by Geometria, New Zealand
Fig.79: Susan Barr
Fig.80: Trygve Aas
Fig.81: Norwegian Maritime Authority
Fig.82: Susan Barr

图书在版编目（CIP）数据

极地文化遗产：不容有失的宝藏：汉英对照／（英）苏珊·巴尔（Susan Barr）著；苏平译．—上海：上海科技教育出版社，2019.5（2022.6重印）

书名原文：Polar Cultural Heritage: Too Important to Lose

ISBN 978-7-5428-6987-6

Ⅰ.①极… Ⅱ.①苏…②苏… Ⅲ.①北极－文化遗产－汉、英 Ⅳ.① P941.62

中国版本图书馆 CIP 数据核字（2019）第 070103 号

责任编辑 王乔琦 伍慧玲
装帧设计 李梦雪
地图由中华地图学社提供，地图著作权归中华地图学社所有

极地科学丛书

极地文化遗产——不容有失的宝藏

Polar Cultural Heritage : Too Important to Lose

苏珊·巴尔（Susan Barr） 著

苏平 译

出版发行		上海科技教育出版社有限公司
		（上海市闵行区号景路159弄A座8楼 邮政编码201101）
网	址	www.sste.com　www.ewen.co
经	销	各地新华书店
印	刷	天津旭丰源印刷有限公司
开	本	720×1000　1/16
印	张	14
版	次	2019年5月第1版
印	次	2022年6月第2次印刷
审 图 号		GS（2019）1692号
书	号	ISBN 978-7-5428-6987-6/N·1057
定	价	80.00元